ROUTLEDGE LIBRARY

20TH CENTURY SCI

Volume 19

EINSTEIN'S THEORY OF UNIFIED FIELDS

EINSTEIN'S THEORY OF UNIFIED FIELDS

M. A. TONNELAT

Translated from the French by
RICHARD AKERIB

Routledge
Taylor & Francis Group
LONDON AND NEW YORK

First published in English in 1966
Second edition published in 1982
Published as *La Théorie du champ unifié d'Einstein et quelques-uns de ses développements* in French.
This edition published in 2014
by Routledge
2 Park Square, Milton Park, Abingdon, Oxfordshire OX14 4RN
and by Routledge
711 Third Avenue, New York, NY 10017

Routledge is an imprint of the Taylor and Francis Group, an informa business

First issued in paperback 2015

British Library Cataloguing in Publication Data
A catalogue record for this book is available from the British Library

ISBN 978-0-415-73519-3 (Set)
eISBN 978-1-315-77941-6 (Set)
ISBN 978-1-138-01362-9 (hbk) (Volume 19)
ISBN 978-1-138-96851-6 (pbk) (Volume 19)
ISBN 978-1-315-77921-8 (ebk) (Volume 19)

Publisher's Note
The publisher has gone to great lengths to ensure the quality of this book but points out that some imperfections from the original may be apparent.

Disclaimer
The publisher has made every effort to trace copyright holders and would welcome correspondence from those they have been unable to trace.

EINSTEIN'S THEORY OF UNIFIED FIELDS

M. A. Tonnelat

Maître de Conférences à la Sorbonne

With a Preface by André Lichnerowicz

Translated from the French by

Richard Akerib

GORDON AND BREACH Science Publishers

New York London Paris

Gordon and Breach, Science Publishers, Inc.
One Park Avenue
New York, NY 10016

Gordon and Breach Science Publishers Ltd.
42 William IV Street
London WC2N 4DE

Gordon & Breach
58, rue Lhomond
75005 Paris

First Published September 1966
Second Printing February 1982

French edition originally published as *La Théorie du champ unifié d'Einstein et quelques-uns de ses développements*. Copyright © Gauthier-Villars, 1955.

Library of Congress Catalog Card Number: 65-16831. ISBN 0 677 00810 4.

Preface to the French Edition

Some time ago, in a small office at the Henri Poincaré Institute, I met Mrs. Tonnelat. We used to discuss our views on fields and sources within the frame of classical relativistic theory. At the time, it did not enter my mind that I would some day have the privilege of presenting the following book. Its author was already exhibiting a highly synthesizing mind and an intellectual honesty which make this book.

As a legacy, Einstein has left an enigmatic theory that scientists view with suspicion and hope. A large amount of work is necessary to either prove or disprove the theory. Even if, as is probable, this theory does not yield the final understanding of physical fields, the work will still be fruitful in that it will lead to a better understanding of the unsuccessful geometrization of the fields and of the ambition itself.

This book does not pretend to be a complete treatise on the theory. We are, by no means, at the stage where such a treatise is necessary. However, the book may prove to be a good research tool to further work on the theory.

Cutting across the multitude of original papers,

Mrs. Tonnelat has succeeded in synthesizing and criticizing the various points of view. She starts with an exposition of the principles of the theory and analyzes the nature of the geometrical synthesis of an affine connection and a fundamental tensor of rank two which satisfies the field equations derived from a variational principle. These equations exhibit the necessary geometrical and physical invariance. The first sides of these equations satisfy also conservation identities whose importance has been exhibited in general relativity.

The difficulties begin when one has to interpret the geometry to find, within the theory, gravitation and electromagnetism and to compare these equations with those of general relativity. Two approaches are available: to find rigorous and as simple as possible solutions of the field equations or to construct approximate solutions. With respect to the first method, the spherically symmetric solutions are the simplest and the most amenable to interpretation and discussion. With respect to the second approach, the manner of approximating is equivocal and implies an a priori interpretation. This labor was started all over the world and it was quite difficult to obtain an idea of the total work carried out with various methods, ideas and implications.

Mrs. Tonnelat's book put in evidence the considerable difficulties involved. Appendices detail certain delicate calculations and discuss the isothermic coordinates which seem, to Mrs. Tonnelat and myself, most important for the future developments of the theory.

It was up to the scientist who, in addition to her other achievements, was the first to explicitly solve the equations tying the affine connection to the fundamental tensor to give us this book.

André Lichnerowicz

Acknowledgments

"God is sophisticated but not malicious" —Albert Einstein

I wanted to assemble here the principles and some of the developments of the Einstein-Schrödinger Unified Field Theory. This is not a complete exposition of the theory. Despite their intrinsic interest, I have systematically left out certain original work of which the point of view would have detracted form the homogeneity of this book. Notably among these researches are those referring to Cauchy's problem treated in previous texts.

In a large measure, I have unified the notations used by various authors. However, in some particular cases, I have adhered to the original notation so as to facilitate reference to the papers.

Most of the chapters arrive at conclusions which raise more or less important difficulties. This is the common lot of all the theories. The unified field theory continues the simplicity of its principles with a profusion of calculations and a wealth of formalism. It is thus difficult to surmount the mathematical complications and to decide between the various physical interpretations to enable one to compare the hopes and realizations of the theory.

One must not seek here a didactic exposition of the acquired results but the more or less happy and complete development of a theory in the process of formation. This book is just a collection of work whose unique goal is to facilitate research on the subject. The fragmentary conclusions which are reached can be only headings of chapters for further work.

M. A. Tonnelat

Contents

Introduction

PURPOSE AND METHODS OF UNIFIED THEORIES

1. FIELD AND CHARGES IN THE GENERAL THEORY OF RELATIVITY.

In developing the General Theory of Relativity, Einstein succeeded in giving a purely geometrical interpretation of gravitation. Newton's law of gravitation was replaced by six independent conditions imposed on the structure of the universe and the trajectories of material particles became geodesics in a four dimensional Riemann space.

This Riemannian space which serves as a vehicle for the general theory is completely determined by the line element: *

$$ds^2 = g_{\mu\nu}\, dx^{\mu}\, dx^{\nu}. \qquad (1)$$

The affine connection between Riemannian spaces is well defined in terms of the metric tensor $g_{\mu\nu}$. It is equal to the Christoffel symbols:

* We use here the usual summation rule that any repeated index indicates a sum over the index.

$$\left\{{\rho \atop \mu\nu}\right\} = g^{\rho\sigma}\{\partial_\mu\, g_{\sigma\nu} + \partial_\nu\, g_{\mu\sigma} - \partial_\sigma\, g_{\mu\nu}\},\quad \partial_\mu = \frac{\partial}{\partial x^\mu} \tag{2}$$

The structure of a Riemannian space is completely determined. It differs from the euclidean structure by the existence of a curvature which is given by the Riemann-Christoffel tensor.

$$G^{\rho}_{\mu\nu\sigma} = \partial_\sigma\left\{{\rho \atop \mu\nu}\right\} - \partial_\nu\left\{{\rho \atop \mu\sigma}\right\} + \left\{{\lambda \atop \mu\nu}\right\}\left\{{\rho \atop \lambda\sigma}\right\} - \left\{{\lambda \atop \mu\sigma}\right\}\left\{{\rho \atop \lambda\nu}\right\} \tag{3}$$

and the curvature is thus defined in terms of the $g_{\mu\nu}$.

1°. In empty space (exterior case), one can obtain the field equations by imposing the following conditions on the space:

$$S_{\mu\nu} = 0, \tag{4}$$

where $S_{\mu\nu}$ is a tensor of rank two subject to being a function of the $g_{\mu\nu}$ and its first two derivatives. The ten equations given by (4) are not all independent since theoretically at least, they would then determine completely the metric tensor and would restrict the choice of the reference system. There can therefore be no more than six independent conditions between the ten components of $S_{\mu\nu}$ to enable us to choose freely the coordinate system in a four dimensional space. We will therefore assume that the $S_{\mu\nu}$ must satisfy the four conservation equations

$$\nabla_\rho S_\mu{}^\rho = 0. \tag{5}$$

∇_ρ is the covariant gradient formed with the aid of the Christoffel symbols. We thus have four arbitrary conditions in the determination of the metric tensor.

E. Cartan (2) has shown that the only tensor $S_{\mu\nu}$ satis-the above conditions must have the form

$$S_{\mu\nu} = h\left\{G_{\mu\nu} - \frac{1}{2}\, g_{\mu\nu}(G - 2\lambda)\right\}, \tag{6}$$

h and λ being constants and $G_{\mu\nu}$ being the contracted Riemann tensor known as the Ricci tensor:

$$G_{\mu\nu} = G^{\rho}_{\mu\nu\rho} = \partial_\rho\left\{{\rho \atop \mu\nu}\right\} - \partial_\nu\left\{{\rho \atop \nu\rho}\right\} + \left\{{\lambda \atop \mu\nu}\right\}\left\{{\rho \atop \lambda\rho}\right\} - \{\rho\} - \left\{{\lambda \atop \mu\rho}\right\}\left\{{\rho \atop \lambda\nu}\right\} \tag{7}$$

If, from (4) and (6), we form the invariant

$$S = g^{\mu\nu} S_{\mu\nu} = 0$$

we obtain

$$G = g^{\mu\nu} G_{\mu\nu} = 4\lambda. \tag{8}$$

Thus, in vacuum, the condition (4) is equivalent to

$$G_{\mu\nu} = \lambda g_{\mu\nu}. \tag{9}$$

2°. In matter or in the presence of an electromagnetic field (interior case) one must balance the effects of the conservative tensor $S_{\mu\nu}$ by those of another conservative tensor: this tensor will represent the energy-momentum of the field or the matter. We will then have

$$S_{\mu\nu} = \chi T_{\mu\nu} \text{ with } \nabla_\rho T_\mu{}^\rho = 0, \tag{10}$$

χ being a constant. For h = 1, we have

$$G_{\mu\nu} - \frac{1}{2} g_{\mu\nu}(G - 2\lambda) = \chi T_{\mu\nu} \tag{11}$$

The above relation has a particular character, for while the interpretation of $S_{\mu\nu}$ is a purely geometrical one, that of $T_{\mu\nu}$ is not. Further $T_{\mu\nu}$ contains:

a. The energy momentum contribution of the electromagnetic field which in Maxwell's theory is represented by

$$\tau_{\mu\nu} = \frac{1}{4} g_{\mu\nu} F_{\rho\sigma} F^{\rho\sigma} - F_{\mu\rho} F^{\rho}_{\nu} . \tag{12}$$

b. The contribution of a distribution of matter of density μ.

$$M_{\rho\sigma} = \mu u_\rho u_\sigma, \tag{13}$$

u_ρ being the four-vector velocity of space:

$$u^\mu = \frac{dx^\mu}{ds}, \qquad u^\mu u_\mu = 1. \tag{14}$$

If we assume the simultaneous presence of charged material particles and an electromagnetic field, the condition

$$\nabla_\rho T_\mu{}^\rho = 0 \quad \text{or} \quad \nabla_\rho M_\mu{}^\rho = - \nabla_\rho \tau_\mu{}^\rho \tag{15}$$

becomes

$$\nabla_\lambda(\mu u_\sigma u^\lambda) = - F_{\sigma\lambda} J^\lambda \tag{16}$$

if $\tau_{\mu\nu}$ is the Maxwell tensor as expressed in (12) and

$$J^\lambda = \nabla_\rho F^{\lambda\rho}. \tag{17}$$

Equation (16) is identical to the first group of Maxwell's equations if we assume that J^λ is the four vector current. If this is true and if we accept Lorentz's hypothesis that all currents are convection currents, then we always have:

$$J^\lambda = \rho u^\lambda \tag{18}$$

ρ being the charge density. If we multiply (16) by u^σ and sum over σ taking into account (14), we have

$$\nabla_\lambda(\mu u^\lambda) = 0. \tag{19}$$

This is just the continuity equation; (16) can then be written as

$$u^\lambda \nabla_\lambda u_\sigma = -\frac{\rho}{\mu} F_{\sigma\lambda} u^\lambda. \tag{20}$$

1°. The trajectory of an uncharged material particle ($p = 0$) will be

$$u^\lambda \nabla_\lambda u_\sigma = 0 \tag{21}$$

which can be rewritten in view of (14), in the form:

$$\frac{d^2 x^\rho}{d s^2} + \left\{ {\rho \atop \mu\nu} \right\} \frac{dx^\mu}{ds} \frac{dx^\nu}{ds} = 0$$

and can also be obtained from the variational principle

$$\delta \int ds = 0.$$

This is the equation for a geodesic in a Riemannian space defined by the line element ds^2 given in (1): thus the trajectories of material particles coincide with the geodesics of a Riemannian space.

2°. In the case of a charged particle ($\rho \neq 0$), Eq. (20) shows that the trajectories differ from the geodesics of a Riemannian space. They could be interpreted as geodesics of a Finsler space in which the metric tensor is determined from the element

$$ds' = (g_{\mu\nu}\, dx^{\mu}\, dx^{\nu})^{\frac{1}{2}} + \frac{e}{m}\, \varphi_{\mu}\, dx^{\mu},$$

by putting $e/m = \rho/\mu$ and assuming that φ_{μ} is the four vector electromagnetic potential such that $F_{\mu\nu} = \partial_{\mu}\varphi_{\nu} - \partial_{\nu}\varphi_{\mu}$. The ratio e/m varies from one particle to another. We must therefore associate with each particle a particular Finsler space whose geodesics will be the trajectories.

2. ROLE AND POSSIBILITIES OF UNIFIED THEORIES.

General Relativity separates quite radically the gravitational field from

a. the electromagnetic field which has not had any geometrical interpretation and forms, with the gravitational field, a heterogenous ensemble.

b. the sources of field which conserve a phenomenological interpretation even when one talks about uncharged particles.

The so-called unified theories attempt to remove the heterogenous features of the combined gravitation and electromagnetic fields.

Other theories, neglecting this problem, have attacked the duality of fields and particles in the purely electromagnetic domain. Such are, for example, the theories of Mie and Born-Infeld (122). Their theories have not attempted a geometrical interpretation of the fields. Despite this, they have often been labeled unified theories. To remove any confusion in the use of this term, we shall refer to them as non-dual theories.

If we were to take into account all the new problems which would arise in the integration of material tensors in a purely geometrical scheme, it would seem that the difficulties, which would be encountered by a non-dual unified theory, would be insurmountable. It was Einstein's ambition to resolve these difficulties. This, perhaps, is not impossible, since, quite often, a question properly asked would remove in one step all the difficulties which, separately, are irreducible.

We shall say that a unified theory is one which will unite the electromagnetic and the gravitational fields into a single hyperfield whose equations are the conditions imposed on the geometrical structure of the universe.

1°. The equations which will describe the behavior of the gravitational and the electromagnetic field in empty space are, by definition, relative to a unified "exterior" case and are always written as

$$S_{\mu\nu} = 0 \quad \text{or} \quad R_{\mu\nu} = 0$$

$R_{\mu\nu}$ being the Ricci tensor formed with some given affine connection. These equations correspond to the "interior" case of gravitation and electromagnetism proposed by the general theory.

2°. In the presence of matter, the equations of a unified theory will include an additional term since we are not, in general, attempting to integrate the field with its sources.

To impose on the structure of the universe the additional conditions that will lead to the electromagnetic field, we cannot stay in purely Riemannian space. The description of such a space is complete when the metric tensor is given. The conditions imposed on the curvature tensor in such a space are sufficient to lead to the interpretation of the gravitational field. The electromagnetic field cannot be described in it.

Additional conditions on the curvature can be obtained by going to a more complex space than the one used by Einstein for General Relativity. This can be done in one of two ways:

a. by increasing the number of dimensions of a Riemannian space.

b. by going into a more general variety of spaces with any affine connection which will give us more latitude in the definition of the parallel displacement of a vector along an infinitely small closed contour.

3. UNIFIED THEORIES WITH MORE THAN FOUR DIMENSIONS.

The theories with (in spaces of) more than four dimensions occupy an important place in the development of unitary theories. The first of these was advanced by Kaluza in 1921 (1), (10), (12). This was further developed by Einstein and Mayer (1), (12), in 1932. A recent modification of the basic postulates has led to theories with 15 field variables by Jordan (1947) and Thiry (1948) (10), (12). All these theories are developed in a five-dimensional space whose geometry leads to a formalism which is convenient for the interpretation of the laws of the generalized fields. For this reason, these theories are all subject to a cylindrical condition which displays clearly the peculiarity of the fifth dimension. Only those coordinate transformations, which satisfy the cylindrical condition, will lead to covariant field equations.

A further development of the five-dimensional formalism has been carried out by the "projective" theories.

The cylindrical condition imposed on these theories can be interpreted in a natural way as a projective condition and shows more clearly the purely auxiliary role of the five-dimensional space.

Along with these theories, other developments have been carried out in a different way. To circumvent the introduction of a pure formalism, one can assume that the physical space actually is five-dimensional. The difference between this postulate and the above (i. e., between a geometry and a geometrical formalism) lies in the interpretation of the cylindrical condition. The condition becomes a structural hypothesis which must be satisfied by the five-dimensional physical space. The Einstein-Bergmann-Bargman theory (1), (12), assumes that the five-dimensional space is closed by the coordinate x^5. The six-dimensional theory of Podolanski (12) gives the space a sheet structure. These assumptions, although gratuitous, are nearer to the spirit of General Relativity which lead to a geometrical character of gravitation.

The great advantage of five-dimensional theories, or at least of some of them, is the following: in a four-dimensional Riemann space, the trajectories of charged particles are not geodesics; it is possible to interpret these trajectories as geodesics in a Finsler space under the condition that for each type of particle characterized by the ratio e/m, a different Finsler space is used. It is possible to show however that one can obtain a parametric representation in a five-dimensional space corresponding to a family of trajectories relative to a given e/m: these trajectories then coincide with the geodesics of a five-dimensional Riemann space.

This remarkable result would push us to develop a unified theory in a five-dimensional formalism. Einstein's and Maxwell's equations can be brought into a satisfactory unified formalism. Aside from criticisms particular to each theory, a general criticism that can be applied to five-dimensional theories is the introduction of an additional condition, namely, the necessity of using the cylindrical condition. In particular the criticism that has been leveled against them is that they lead to a simple codifying in a five-dimensional formalism of the equations of Einstein and Maxwell. No matter how interesting such a synthesis is, it should lead, as General Relativity did, to predictions that will confirm or reject the theory.

It is in answer to these objections that the Jordan-Thiry theory with 15 field variables was developed. Essentially, the consequences of these theories are:

1. The gravitational constant χ becomes a weakly varying factor, the variations being a function of the variations of e/m.

2. The laws of Einstein and Mawell introduce additional terms due to the variation of χ. If χ is constant, one reverts to the classical laws.

3. There is a fifteenth field equation relative to the variations of χ which implies that in the absence of any charge ($\rho = 0$), the presence of matter ($\mu \neq 0$) leads to the creation of a magnetic field. In this way we arrive at the prediction of the existence of a field due to matter in motion and particularly for a rotating body (Blackett effect).

4. FOUR-DIMENSIONAL UNIFIED THEORIES. SPACES WITH GENERAL AFFINE CONNECTIONS.

Along with five-dimensional theories, other theories, since 1918, have attempted the synthesis of gravitation and electromagnetism in the context of a four-dimensional space-time. They must therefore start from more general multiplicities than the Riemannian variety in order to enable one to impose supplementary structural conditions that will coincide with or better modify, the classical electromagnetic equations.

The Riemannian space, which is described by a single type of curvature, will be replaced by a more general continuum formed by considering a space with a more general affine connection. Cartan (3) has shown that the structure of such a space defined by the coefficients $\Gamma_{\mu\nu}{}^{\rho}$ will, in general, have

a. A rotational curvature defined by the tensor Ω_{μ}^{ρ} which generalizes the Riemann-Christoffel tensor*

$$\Omega_{\mu}^{\rho} = -\frac{1}{2} R_{\mu\nu\sigma}^{\rho} \, [dx^{\nu} \, \delta x^{\sigma}]$$

$$= -\frac{1}{2}\{\partial_{\sigma}\Gamma_{\mu\nu}^{\rho} - \partial_{\nu}\Gamma_{\mu\sigma}^{\rho} + \Gamma_{\mu\nu}^{\lambda}\,\Gamma_{\lambda\sigma}^{\rho} - \Gamma_{\mu\sigma}^{\lambda}\,\Gamma_{\lambda\nu}^{\rho}\} \, [dx^{\nu} \, \delta x^{\sigma}] \, .$$

---

* The $[dx^{\mu}\,\delta x^{\nu}]$ denote the differential increments and are:

$$[dx^{\mu}\,\delta x^{\nu}] = dx^{\mu}\,\delta x^{\nu} - \delta x^{\mu}\,dx^{\nu}.$$

b. A (homothetic) curvature which suffices to characterize the invariant $\Omega = \Omega^{\mu}_{\mu}$ and which is zero in a Riemannian space.

c. A torsion which is zero in a Riemann space and is given by

$$\Omega^{\rho} = - \Gamma^{\rho}_{\mu\nu} \, [dx^{\mu} \, \delta x^{\nu}].$$

Most authors did not immediately perceive the nature of the elements a, b and c at their disposal. However, any affine four-dimensional unified theory assumes the existence of one or several of these elements. All these theories, although apparently different, can be classified in three distinct categories depending on whether they assume the elements a-b, c, a-b-c, which are used to characterize a Riemannian space and develop the general theory of relativity.

The first theory of this type was developed by Weyl (13). He assumed that a parallel displacement does not only modify the direction of a vector but also its magnitude. This assumption is equivalent to postulating the non-vanishing of the homothetic curvature and the conservation of elements a and b. Analytically, this assumption introduces gauge transformation in addition to coordinate transformations. The affine connection is then expressed in terms of the metric tensor and a four vector φ_{μ} tied to the gauge transformations. Thus, to determine the structure of the universe and the field equations, it is necessary to use the ensemble $(g_{\mu\nu}, \varphi_{\mu})$ and not simply ds^2.

Weyl's theory does not modify the symmetrical character of the space: this is equivalent to assuming a non-torsional space. The torsion will lead to a non-zero value for a closed infinitely small contour. The torsional spaces introduced by Cartan (3) were used in unitary theories by Eyrand (1926) (12), by Infeld (1928) (12), and finally by Einstein (5), (12), (125). A theory developed by Einstein (12) has even been based on the existence of a strictly torsional space which permits the definition of absolute parallelism. In many of their features, the theories of Eyrand (a, b, c) and Infeld (a, c) and Einstein's theory (1923) (5), (125) are forerunners of Einstein's unified theory which we shall describe. Unfortunately, they have unnecessarily limited the formalism at their disposal and did not succeed in exploiting the richness of the structure of a general affine connection to derive new laws.

5. EINSTEIN'S THEORY.

Einstein's theory starts from a general formalism which, from the beginning, conserves without restricting properties a, b and c. If a four-dimensional unitary theory will give satisfactory results starting from a variational principle, it will be within the context of this theory. As we shall see the price that has been paid in return for the completeness of the theory lies in the ambiguities at its foundations. A serious difficulty arises in attempting to resolve these ambiguities and to choose, in a consistent way, the contracted curvature tensor, the metric tensor and the fields. The theory has the advantage of not restricting the possibilities offered by any affine theory.

We stated earlier that the often justified criticism leveled at unified theories was that they restricted themselves to synthesizing existing and well-established laws. However, although the theoretical interest in a geometrical synthesis of gravitation and electromagnetism is undeniable it is nevertheless true that it is rather useless to recover Einstein's and Maxwell's laws in theories that introduce a complicated formalism. One would wish to go further.

Einstein's theory opens new perspectives for us. These are:

1. The laws of Einstein and Maxwell are modified. They contain additional terms which represent the interaction of the gravitational and the electromagnetic fields. It seems thus that one can justify the creation of a magnetic field by a purely material distribution.

2. In the general case, this type of effect is difficult to bring out due to the complicated nature of the unified

field equations. In the particular case of spherical or cylindrical symmetry, one can obtain rigorous laws for this particular case. One can thus determine precisely the laws obeyed by the field of a pulsating or rotating body. This would permit a justification or a modification of Blackett's empirical formula.

3. The laws of electromagnetism which result from Einstein's theory are not linear. They seem to predict, theoretically at least, some new effects, such as the scattering of light by light and the possibility of defining a field in the center of a particle.

4. The relations between inductions and fields in Einstein's theory (or at least between certain quantities tied to inductions and fields) have the same form as in Born's theory. In the case of a static spherically symmetric solution, one can show the existence of a field which is finite in the center of the particle.

One would think that Einstein's hope to integrate with the field its singularities and the sources of the field in a geometrical synthesis can be realized. It is difficult to be so positive about this hope. Assuming that even this first objective has been attained, it is difficult to see the next step which will lead the theory to a reinterpretation of quantum theories, a step which Born's theory was unable to bridge in a satisfactory way. Perhaps as we stated earlier, the problem considered as a whole, namely unifying the fields and unifying fields and sources, will be more easily solved if stated more judiciously.

In an immediate way, the other difficulties of the theory result on the one hand from the richness of the formalism and on the other hand from the complexity of the solutions. On the one hand, one encounters ambiguities which must be resolved, interpretations which are not always clear: the choice of the tensor $R_{\mu\nu}$ used in the variational principle, the "true" metric, the determination of the energy-momenttum tensor and the derivation of the geodesics. All these points raise problems, a few of which are on the way to being solved. On the other hand, one must handle complicated solutions which, except in particular cases and sometimes even in these, hide the physical contents that one could find.

Despite this, a number of satisfactory results have already been obtained. Others need to be studied further. Perhaps in this fashion, we will be able to proceed toward a clearer conception of the nature of the sources of a field and from this to the scope of the theory. One can detect in both Einstein and Schrödinger a mixture of discouragement and great hopes in reference to this point.

Be that as it may, Einstein's theory unites the realization of a satisfactory synthesis, obtained from a general principle, to the possibility of new predictions in the classical domain. Such are almost always the signs which characterize a fruitful and important physical theory.

1

Mathematical Introduction

A. RELATIONS BETWEEN THE SYMMETRICAL AND ANTISYMMETRICAL PARTS OF THE TENSORS $g_{\mu\nu}$ and $g^{\mu\nu}$ (54), (57).

1. DEFINITIONS.

Consider a tensor of rank two $g_{\mu\nu}$ whose determinant is g and let $gg^{\mu\nu}$ be the minor relative to each element $g_{\mu\nu}$. We have by definition:

$$g_{\mu\rho}\, g^{\mu\sigma} = g_{\rho\mu} g^{\sigma\mu} = \delta_\rho^\sigma \tag{1. 1}$$

$$dg = gg^{\mu\nu}\, dg_{\mu\nu} = -\, gg_{\mu\nu}\, dg^{\mu\nu} \tag{1. 2}$$

whence we have

$$dg_{\mu\nu} = -\, g_{\rho\nu}\, g_{\mu\sigma}\, dg^{\rho\sigma}, \quad dg^{\mu\nu} = -\, g^{\rho\nu}\, g^{\mu\sigma}\, dg_{\rho\sigma}. \tag{1. 3}$$

Let us now write $g_{\mu\nu}$ and $g^{\mu\nu}$ in terms of their symmetric parts $\gamma_{\mu\nu}$ and $h^{\mu\nu}$ and antisymmetric parts $\varphi_{\mu\nu}$ and $f^{\mu\nu}$, respectively:

$$g_{\mu\nu} = \gamma_{\mu\nu} + \varphi_{\mu\nu} \tag{1. 4}$$

$$g^{\mu\nu} = h^{\mu\nu} + f^{\mu\nu}. \tag{1.5}$$

We will now adopt the following convention: the same letter will be used to denote the covariant components of a tensor, its determinant and the minor relative to each of its element; thus γ and φ will denote the determinant formed by the elements $\gamma_{\mu\nu}$ and $\varphi_{\mu\nu}$ and $\gamma\gamma^{\mu\nu}$ and $\varphi\varphi^{\mu\nu}$ the minors relative to these elements. In an analogous way h and f will be the determinants formed by the elements $h_{\mu\nu}$ and $f_{\mu\nu}$ and $hh^{\mu\nu}$ and $ff^{\mu\nu}$ will be the minors relative to these elements.

We always have:

$$\gamma_{\mu\rho}\gamma^{\mu\sigma} = \delta^{\sigma}_{\rho}, \quad \varphi_{\mu\rho}\,\varphi^{\mu\sigma} = \delta^{\sigma}_{\rho}, \quad h_{\mu\rho}\,h^{\mu\sigma} = \delta^{\sigma}_{\rho}, \qquad (1.6)$$

$$f_{\mu\rho}\,f^{\mu\sigma} = \delta^{\sigma}_{\rho};$$

$$d\gamma = \gamma\gamma^{\mu\nu}\,d\gamma_{\mu\nu}, \qquad d\varphi = \varphi\varphi^{\mu\nu}\,d\varphi_{\mu\nu}$$

$$dh = hh^{\mu\nu}\,dh_{\mu\nu}, \qquad df = ff^{\mu\nu}\,df_{\mu\nu}. \qquad (1.7)$$

Let $\epsilon_{\mu\nu\rho\sigma} = \epsilon^{\mu\nu\rho\sigma}$ be the Levi-Civita tensor (equal to ± 1 as to whether we have an even or odd permutation with respect to the indices and zero if any two indices are equal). The relations between the covariant and the contravariant components of an antisymmetric tensor are:

$$\varphi_{\mu\nu} = \frac{\sqrt{\varphi}}{2}\,\epsilon_{\mu\nu\rho\sigma}\,\varphi^{\rho\sigma}, \qquad \varphi^{\mu\nu} = \frac{1}{2\sqrt{\varphi}}\,\epsilon^{\mu\nu\rho\sigma}\varphi_{\rho\sigma}, \qquad (1.8)$$

$$f_{\mu\nu} = \frac{\sqrt{f}}{2}\,\epsilon_{\mu\nu\rho\sigma}\,f^{\rho\sigma}, \qquad f^{\mu\nu} = \frac{1}{2\sqrt{f}}\,\epsilon^{\mu\nu\rho\sigma}f_{\rho\sigma}, \qquad (1.9)$$

whence the determinants φ and f are:

$$\sqrt{\varphi} = \frac{1}{8}\,\epsilon^{\mu\nu\rho\sigma}\,\varphi_{\mu\nu}\,\varphi_{\rho\sigma} = (\varphi_{12}\,\varphi_{34} + \varphi_{31}\varphi_{24} + \varphi_{23}\,\varphi_{14}) \qquad (1.10)$$

$$\sqrt{f} = \frac{1}{8}\,\epsilon^{\mu\nu\rho\sigma}\,f_{\mu\nu}\,f_{\rho\sigma} = (f_{12}\,f_{34} + f_{31}\,f_{24} + f_{23}\,f_{14}) \qquad (1.11)$$

For a symmetric tensor of rank two, we will have:*

$$\gamma \epsilon_{\mu\nu\rho\sigma} \gamma^{\mu\lambda} \gamma^{\nu\tau} = \epsilon^{\mu\nu\lambda\tau} \gamma_{\mu\rho} \gamma_{\nu\sigma} \tag{1.12}$$

$$h \epsilon_{\mu\nu\rho\sigma} h^{\mu\lambda} h^{\nu\tau} = \epsilon^{\mu\nu\lambda\tau} h_{\mu\rho} h_{\nu\sigma} \tag{1.13}$$

We will always assume that g, γ and h are different from zero. Thus, we can always compute $\gamma^{\mu\nu}$ and $h^{\mu\nu}$ starting from $\gamma_{\mu\nu}$ and $h_{\mu\nu}$. If φ and 1/f are zero, the definitions of $\varphi^{\mu\nu}$ and $f_{\mu\nu}$ as given in (1. 8) and (1.9) have no meaning. In this case, we shall consider the expression $\sqrt{\varphi}\ \varphi^{\mu\nu}$ and $f_{\mu\nu}/\sqrt{f}$ which will always have a meaning since they represent $\frac{1}{2}\epsilon^{\mu\nu\rho\sigma} \varphi_{\rho\sigma}$ and $\frac{1}{2}\epsilon_{\mu\nu\rho\sigma} f^{\rho\sigma}$.

2. FUNDAMENTAL RELATIONS.

The determinant g can be expressed in terms of the symmetric and antisymmetric determinants γ and φ. In the same fashion, one can express the minor $gg^{\mu\nu}$ in terms of $\gamma\gamma^{\mu\nu}$ and $\varphi\varphi^{\mu\nu}$. Thus one can write:

$$g = \gamma + \varphi + \frac{\gamma}{2} \gamma^{\mu\rho} \gamma^{\nu\sigma} \varphi_{\mu\nu} \varphi_{\rho\sigma}, \tag{1.14}$$

$$gg^{\mu\nu} = \gamma\gamma^{\mu\nu} + \varphi\varphi^{\mu\nu} + \gamma\gamma^{\mu\rho} \gamma^{\nu\sigma} \varphi_{\rho\sigma} + \varphi\varphi^{\mu\rho} \varphi^{\nu\sigma} \gamma_{\rho\sigma}. \tag{1.15}$$

In a reciprocal way 1/g, 1/h, 1/f are the determinants of $g^{\mu\nu}$, $h^{\mu\nu}$ and $f^{\mu\nu}$. We will thus have relations analogous to (1. 14) and (1. 15) between 1/g, 1/h and 1/f:

$$\frac{1}{g} = \frac{1}{h} + \frac{1}{f} + \frac{1}{2h} h_{\mu\rho} h_{\nu\sigma} f^{\mu\nu} f^{\rho\sigma}, \tag{1.16}$$

$$\frac{1}{g} g^{\mu\nu} = \frac{1}{h} h^{\mu\nu} + \frac{1}{f} f^{\mu\nu} + \frac{1}{h} h_{\mu\rho} h_{\nu\sigma} f^{\rho\sigma} + \frac{1}{f} f_{\mu\rho} f_{\nu\sigma} h^{\rho\sigma}. \tag{1.17}$$

---

* We note here a typographical error in the corresponding formulas (12) of article (57).

Let us separate (1. 15) and (1. 17) into their symmetric and antisymmetric parts. This leads to (1. 18) and (1. 19). One can also obtain the reciprocal relations for (1. 18) and (1. 19). (See Appendix I.) This leads to (1. 20) and (1. 21). These points are summarized in the following table (54), (57):

Symmetric part (s)	Antisymmetric part (a)	
$h^{\mu\nu} = \frac{\gamma}{g}\gamma^{\mu\nu} + \frac{\varphi}{g}\varphi^{\mu\rho}\varphi^{\nu\sigma}\gamma_{\rho\sigma}$	$f^{\mu\nu} = \frac{\varphi}{g}\varphi^{\mu\nu} + \frac{\gamma}{g}\gamma^{\mu\rho}\gamma^{\nu\sigma}\varphi_{\rho\sigma}$	(1. 18)
$\gamma_{\mu\nu} = \frac{g}{h}h_{\mu\nu} + \frac{g}{f}f_{\mu\rho}f_{\nu\sigma}h^{\rho\sigma}$	$\varphi_{\mu\nu} = \frac{g}{f}f_{\mu\nu} + \frac{g}{h}h_{\mu\rho}h_{\nu\sigma}f^{\rho\sigma}$	(1. 19)
$h_{\mu\nu} = \gamma_{\mu\nu} + \varphi_{\mu\rho}\varphi_{\nu\sigma}\gamma^{\rho\sigma}$	$f_{\mu\nu} = \varphi_{\mu\nu} + \gamma_{\mu\rho}\gamma_{\nu\sigma}\varphi^{\rho\sigma}$	(1. 20)
$\gamma^{\mu\nu} = h^{\mu\nu} + f^{\mu\rho}f^{\nu\sigma}h_{\rho\sigma}$	$\varphi^{\mu\nu} = f^{\mu\nu} + h^{\mu\rho}h^{\nu\sigma}f_{\rho\sigma}$	(1. 21)

If we multiply (1. 18s) by (1. 20s) and (1. 18a) by (1. 20a), we obtain the following conditions

$$g^2 = \gamma h = f\varphi \tag{1. 22}$$

We note here the relation (I. 7) (see Appendix I):

$$\varphi\varphi^{\mu\tau}\varphi^{\nu\rho}\gamma_{\mu\nu}\gamma_{\sigma\rho} = \delta^{\tau}_{\sigma}\ (g - \gamma - \varphi) - \gamma\gamma^{\mu\tau}\gamma^{\nu\rho}\varphi_{\mu\nu}\varphi_{\sigma\rho} \tag{1. 23}$$

which reduces for $\sigma = \tau$ to

$$\varphi\varphi^{\mu\sigma}\varphi^{\nu\rho}\gamma_{\mu\nu}\gamma_{\sigma\rho} = \gamma\gamma^{\mu\sigma}\gamma^{\nu\rho}\varphi_{\mu\nu}\varphi_{\sigma\rho} = 2(g - \gamma - \varphi) \tag{1. 24}$$

REMARK (case $\varphi = 0$, $1/f = 0$). (1. 20a) and (1. 21a) in the form given above hold only if $\varphi \neq 0$ and $1/f \neq 0$. If both φ and $1/f$ are zero, one must rewrite (1. 8) and (1. 9) in the more meaningful way

$$2\sqrt{\varphi}\,\varphi^{\mu\nu} = \epsilon^{\mu\nu\rho\sigma}\,\varphi_{\rho\sigma} \qquad \frac{2}{\sqrt{f}}\,f_{\mu\nu} = \epsilon_{\mu\nu\rho\sigma}\,f^{\rho\sigma}. \quad (1.25)$$

From Appendix I, one then obtains:

$$\sqrt{\varphi}\,\varphi^{\mu\nu} = \frac{g}{\sqrt{f}}\,h^{\mu\rho}\,h^{\nu\sigma}\,f_{\rho\sigma} \qquad \frac{1}{\sqrt{f}}\,f_{\mu\nu} = \frac{\sqrt{\varphi}}{g}\,\gamma_{\mu\rho}\gamma_{\nu\sigma}\varphi^{\rho\sigma}. \quad (1.26)$$

These are just (1.20) and (1.21) where we have used $\sqrt{\varphi}\,\varphi^{\mu\nu}$ and $f_{\mu\nu}1/\sqrt{f}$ instead of $\varphi^{\mu\nu}$ and $f_{\mu\nu}$. This is the recipe for obtaining the general forms of equations (1.20) and (1.21). As we shall see later, it is always the products $\sqrt{\varphi}\,\varphi^{\mu\nu}$ and $f_{\mu\nu}/\sqrt{f}$ that appear in the theory.

3. CONJUGATE VARIABLES.

Consider the expressions

$$\mathcal{L} = 2\sqrt{-g} = 2\sqrt{-\gamma}\,L \quad (1.27)$$

with

$$L = \sqrt{\frac{g}{\gamma}} = \sqrt{\frac{h}{g}} = \left\{1 + \frac{\varphi}{\gamma} + \frac{1}{2}\,\gamma^{\mu\rho}\,\gamma^{\nu\sigma}\,\varphi_{\mu\nu}\,\varphi_{\rho\sigma}\right\}^{\frac{1}{2}} \quad (1.28)$$

$$= \left\{1 + \frac{h}{f} + \frac{1}{2}\,h_{\mu\rho}\,h_{\nu\sigma}\,f^{\mu\nu}\,f^{\rho\sigma}\right\}^{\frac{1}{2}}$$

From

$$dg = g g^{\mu\nu}\,dg_{\mu\nu} = -\sqrt{-g}\,g_{\mu\nu}\,d\mathcal{G}^{\mu\nu} \quad (1.29)$$

one can verify that

$$\sqrt{-g}\;g^{\mu\nu} = \frac{\partial\mathcal{L}}{\partial g_{\mu\nu}} \quad \text{and} \quad g_{\mu\nu} = \frac{\partial\mathcal{L}}{\partial(\sqrt{-g}\;g^{\mu\nu})} \quad (1.30)$$

Separation of (1.30) into symmetric and antisymmetric parts leads to:

$$\sqrt{-g}\;h^{\mu\nu} = \frac{1}{2}\left\{\frac{\partial\mathcal{L}}{\partial g_{\mu\nu}} + \frac{\partial\mathcal{L}}{\partial g_{\nu\mu}}\right\}, \quad \sqrt{-g}\;f^{\mu\nu} = \frac{1}{2}\left\{\frac{\partial\mathcal{L}}{\partial g_{\mu\nu}} - \frac{\partial\mathcal{L}}{\partial g_{\nu\mu}}\right\}; \quad (1.31a)$$

$$\gamma_{\mu\nu} = \frac{1}{2}\left\{\frac{\partial\mathcal{L}}{\partial\mathcal{G}^{\mu\nu}} + \frac{\partial\mathcal{L}}{\partial\mathcal{G}^{\nu\mu}}\right\}, \quad \varphi_{\mu\nu} = \frac{1}{2}\left\{\frac{\partial\mathcal{L}}{\partial\mathcal{G}^{\mu\nu}} - \frac{\partial\mathcal{L}}{\partial\mathcal{G}^{\nu\mu}}\right\}. \quad (1.31b)$$

On the other hand, $\mathcal{L}$ is a function of $g_{\mu\nu}$ and one has by definition:

$$d\mathcal{L} = \frac{\partial \mathcal{L}}{\partial g_{\mu\nu}} dg_{\mu\nu} = \frac{1}{2}\left\{\frac{\partial \mathcal{L}}{\partial g_{\mu\nu}} + \frac{\partial \mathcal{L}}{\partial g_{\nu\mu}}\right\} d\gamma_{\mu\nu} + \frac{1}{2}\left\{\frac{\partial \mathcal{L}}{\partial g_{\mu\nu}} - \frac{\partial \mathcal{L}}{\partial g_{\mu\nu}}\right\} d\varphi_{\mu\nu} \tag{1.32a}$$

$$d\mathcal{L} = \frac{\partial \mathcal{L}}{\partial \mathcal{G}^{\mu\nu}} d\mathcal{G}^{\mu\nu} = \frac{1}{2}\left\{\frac{\partial \mathcal{L}}{\partial \mathcal{G}^{\mu\nu}} + \frac{\partial \mathcal{L}}{\partial \mathcal{G}^{\nu\mu}}\right\} d\mathcal{H}^{\mu\nu} + \frac{1}{2}\left\{\frac{\partial \mathcal{L}}{\partial \mathcal{G}^{\mu\nu}} - \frac{\partial \mathcal{L}}{\partial \mathcal{G}^{\nu\mu}}\right\} d\mathfrak{G}^{\mu\nu}, \tag{1.32b}$$

A comparison of (1.31) and (1.32) leads to:*

$$\frac{\partial \mathcal{L}}{\partial \gamma_{\mu\nu}} = \frac{1}{2}\left\{\frac{\partial \mathcal{L}}{\partial g_{\mu\nu}} + \frac{\partial \mathcal{L}}{\partial g_{\nu\mu}}\right\} = \mathcal{H}^{\mu\nu}, \qquad \frac{\partial \mathcal{L}}{\partial \varphi_{\mu\nu}} = \frac{1}{2}\left\{\frac{\partial \mathcal{L}}{\partial g_{\mu\nu}} - \frac{\partial \mathcal{L}}{\partial g_{\nu\mu}}\right\} = \mathfrak{G}^{\mu\nu} \tag{1.33a}$$

$$\frac{\partial \mathcal{L}}{\partial \mathcal{H}^{\mu\nu}} = \frac{1}{2}\left\{\frac{\partial \mathcal{L}}{\partial \mathcal{G}^{\mu\nu}} + \frac{\partial \mathcal{L}}{\partial \mathcal{G}^{\nu\mu}}\right\} = \gamma_{\mu\nu}{}^{(1)}, \qquad \frac{\partial \mathcal{L}}{\partial \mathfrak{G}^{\mu\nu}} = \frac{1}{2}\left\{\frac{\partial \mathcal{L}}{\partial \mathcal{G}^{\mu\nu}} - \frac{\partial \mathcal{L}}{\partial \mathcal{G}^{\nu\mu}}\right\} = \varphi_{\mu\nu} \tag{1.33b}$$

As we have not taken account of the symmetry property of $\gamma_{\mu\nu} = \gamma_{\nu\mu}$ and $\mathcal{H}^{\mu\nu} = \mathcal{H}^{\nu\mu}$ and the antisymmetry property of $\varphi_{\mu\nu} = -\varphi_{\nu\mu}$ and $\mathfrak{G}^{\mu\nu} = -\mathfrak{G}^{\nu\mu}$, the relations (1.33) are quite general. In particular the fields $\mathfrak{G}^{\mu\nu}$ and $\varphi_{\mu\nu}$ are conjugate to each other with respect to the scalar density $\sqrt{-g}$. One can see from the expressions for $h^{\mu\nu}$, $f^{\mu\nu}$, $\gamma_{\mu\nu}$

---

* We note here that the notation $\partial\mathcal{L}/\partial\lambda_{\mu\nu}$ represents the partial derivative of $\mathcal{L}$ taken as if the $\lambda_{\mu\nu}$ or $\lambda_{\nu\mu}$ was an independent variable.

and $\varphi_{\mu\nu}$ as given in (1.18) and (1.19) that the relations (1.33) are correct when $\mathcal{L}$ is given by (1.27). We note also that (1.33) can be written as*

$$f^{\mu\nu} = \frac{2}{\mathcal{L}} \frac{\partial \mathcal{L}}{\partial \varphi_{\mu\nu}}, \qquad \varphi_{\mu\nu} = \frac{2\sqrt{-g}}{\mathcal{L}} \frac{\partial \mathcal{L}}{\partial \mathcal{J}^{\mu\nu}}. \tag{1.34}$$

B. ABSOLUTE DIFFERENTIAL CALCULUS IN A SPACE WITH A GENERAL AFFINE CONNECTION.

4. COVARIANT DERIVATIVES:

For a symmetrical affine connection $\Gamma^{\rho}_{\mu\nu}$, the parallel displacement of a vector can be defined unambiguously by

$$\delta A^{\mu} = - \Gamma^{\mu}_{\underline{\sigma\rho}} A^{\sigma} \delta x^{\rho}. \tag{1.35}$$

On the other hand for a general affine connection $\Gamma^{\rho}_{\mu\nu}$, we can define the parallel displacement by one of the two relations:

$$\delta A^{\mu}_{+} = - \Gamma^{\mu}_{\sigma\rho} A^{\sigma} \delta x^{\rho} \tag{1.36}$$

$$\delta A^{\mu}_{-} = - \Gamma^{\mu}_{\rho\sigma} A^{\sigma} \delta x^{\rho} \tag{1.37}$$

---

* By letting

$$\mathcal{G}^{\mu\nu} = \sqrt{-g}\, g^{\mu\nu} = \frac{1}{\mathcal{G}} \times \text{minor}\, \mathcal{G}_{\mu\nu},$$

$$\mathcal{H}^{\mu\nu} = \sqrt{-g}\, h^{\mu\nu} = \frac{1}{\mathcal{H}} \times \text{minor}\, \mathcal{H}_{\mu\nu}$$

the determinants $\mathcal{G}$, $\mathcal{H}$ and $\mathcal{J}$ of $\mathcal{G}_{\mu\nu}$, $\mathcal{H}_{\mu\nu}$ and $\mathcal{J}_{\mu\nu}$ are

$$\mathcal{G} = \frac{1}{g}, \quad \mathcal{H} = \frac{h}{g^2} = \frac{1}{\gamma} \quad \mathcal{J} = \frac{f}{g^2} = \frac{1}{\varphi}.$$

Whence we have:

$$\mathcal{L} = \frac{2}{\sqrt{-g}} = \frac{2L}{\sqrt{-\mathcal{H}}},$$

$$\varphi_{\mu\nu} = \frac{\partial \mathcal{L}}{\partial \mathcal{J}^{\mu\nu}} = 2 \frac{\sqrt{-g}}{L} \frac{\partial L}{\partial \mathcal{J}^{\mu\nu}} \quad \text{with } L = \left\{1 + \frac{\mathcal{H}}{\mathcal{J}} + \frac{1}{2} \mathcal{H}_{\mu\rho} \mathcal{H}_{\nu\sigma} \mathcal{J}^{\mu\nu} \mathcal{J}^{\nu\sigma}\right\}^{\frac{1}{2}}.$$

which conserve the tensorial character of $\partial_\rho A^\mu + \Gamma^\mu_{\sigma\rho} A^\sigma$. We can thus define two types of covariant derivatives:

$$A^\mu_{+;\rho} = \partial_\rho A^\mu + \Gamma^\mu_{\sigma\rho} A^\sigma \qquad (1.38a)$$

$$A^\mu_{-;\rho} = \partial_\rho A^\mu + \Gamma^\mu_{\rho\sigma} A^\sigma. \qquad (1.38b)$$

In the same fashion, we have:

$$A_{\underset{+}{\mu};\rho} = \partial_\rho A_\mu - \Gamma^\sigma_{\mu\rho} A_\sigma \qquad (1.39a)$$

$$A_{\underset{-}{\mu};\rho} = \partial_\rho A_\mu - \Gamma^\sigma_{\rho\mu} A_\sigma. \qquad (1.39b)$$

If the connection is symmetrical, then (1.38a) coincides with (1.38b) and similarly for (1.39) and we have:

$$A^\mu_{0;\rho} = \partial_\rho A^\mu + \Gamma^\mu_{\underline{\sigma\rho}} A^\sigma \qquad (1.40a)$$

$$A_{\underset{0}{\mu};\rho} = \partial_\rho A_\mu - \Gamma^\sigma_{\underline{\mu\rho}} A_\sigma. \qquad (1.40b)$$

Let $\Gamma^\rho_{\underline{\mu\nu}}$ and $\Gamma^\rho_{\underset{\vee}{\mu\nu}}$ be respectively the symmetrical and antisymmetrical parts of $\Gamma^\rho_{\mu\nu}$:

$$\Gamma^\rho_{\mu\nu} = \Gamma^\rho_{\underline{\mu\nu}} + \Gamma^\rho_{\underset{\vee}{\mu\nu}} \qquad (1.41)$$

If we contract the indices ρ and ν, we have $\Gamma_\mu = \Gamma^\nu_{\underset{\vee}{\mu\nu}}$, Γ_μ being the torsional four vector of the space. The transpose of $\Gamma^\rho_{\mu\nu}$ is defined as

$$\tilde{\Gamma}^\rho_{\mu\nu} = \Gamma^\rho_{\nu\mu} \qquad (1.42)$$

such that

$$\tilde{\Gamma}^\rho_{\mu\nu} = \Gamma^\rho_{\underline{\mu\nu}} - \Gamma^\rho_{\underset{\vee}{\mu\nu}}. \qquad (1.43)$$

5. TENSOR DENSITIES.

From (1. 39) the covariant derivative of a tensor $g_{\mu\nu}$ is:

$$g_{\underset{+\,-}{\mu\nu};\rho} = \partial_\rho g_{\mu\nu} - \Gamma^{\sigma}_{\mu\rho}\, g_{\sigma\nu} - \Gamma^{\sigma}_{\rho\nu}\, g_{\mu\sigma} \tag{1. 44}$$

Multiplying (1. 44) by $g^{\mu\nu}$, we obtain:

$$g^{\mu\nu} g_{\underset{+\,-}{\mu\nu};\rho} = g^{\mu\nu}\partial_\rho g_{\mu\nu} - (\Gamma^{\mu}_{\mu\rho} + \Gamma^{\mu}_{\rho\mu}) = \left\{2\,\frac{\partial_\rho\sqrt{-g}}{\sqrt{-g}} - \Gamma^{\mu}_{\underline{\mu\rho}}\right\}. \tag{1. 45}$$

$\sqrt{-g}$ is a tensor density. Its covariant derivative will then be defined by

$$(\sqrt{-g})_{;\rho} = \partial_\rho\sqrt{-g} - \sqrt{-g}\,\Gamma^{\sigma}_{\underline{\rho\sigma}}\,. \tag{1. 46}$$

We will then have:

$$g^{\mu\nu}\, g_{\underset{+\,-}{\mu\nu};\rho} = \frac{2}{\sqrt{-g}}(\sqrt{-g})_{;\rho}. \tag{1. 47}$$

If we multiply (1. 44) by $g^{\mu\lambda}\, g^{\tau\nu}$, we obtain:

$$g^{\mu\nu}_{+\,-;\rho} = \partial_\rho g^{\mu\nu} + \Gamma^{\mu}_{\sigma\rho}\, g^{\sigma\nu} + \Gamma^{\nu}_{\rho\sigma}\, g^{\mu\sigma}. \tag{1. 48}$$

If we define the tensor density

$$\mathcal{G}^{\mu\nu} = \sqrt{-g}\, g^{\mu\nu}, \tag{1.49}$$

its covariant derivative will be

$$\mathcal{G}^{\mu\nu}_{+\,-;\rho} = \partial_\rho\,\mathcal{G}^{\mu\nu} + \Gamma^{\mu}_{\sigma\rho}\,\mathcal{G}^{\sigma\nu} + \Gamma^{\nu}_{\rho\sigma}\,\mathcal{G}^{\mu\sigma} - \mathcal{G}^{\mu\nu}\,\Gamma^{\sigma}_{\underline{\sigma\rho}} \tag{1. 50}$$

by taking into account (1. 46).

6. FUNDAMENTAL IDENTITIES BETWEEN THE $\mathcal{G}^{\mu\nu}_{\substack{+-};\rho}$ and between the $g_{[\underset{+-}{\mu\nu};\rho]}$.

a. Let us form $\mathcal{G}^{\mu\rho}_{\substack{+-};\rho} - \mathcal{G}^{\rho\mu}_{\substack{+-};\rho}$. By contracting (1.50), we have the identity:

$$\frac{1}{2}\left\{\mathcal{G}^{\mu\rho}_{\substack{+-};\rho} - \mathcal{G}^{\rho\mu}_{\substack{+-};\rho}\right\} \equiv \partial_\rho \mathfrak{g}^{\mu\rho} - \mathfrak{H}^{\mu\rho}\,\Gamma_\rho \tag{1.51}$$

by using (1.5) and the corresponding notations for tensorial densities.

$$\mathcal{G}^{\mu\nu} = \mathfrak{H}^{\mu\nu} + \mathfrak{g}^{\mu\nu} \quad (\mathfrak{H}^{\mu\nu} = \sqrt{-g}\,h^{\mu\nu};\ \mathfrak{g}^{\mu\nu} = \sqrt{-g}\,f^{\mu\nu}). \tag{1.52}$$

b. Let us form the difference between the circular permutation of $g_{\underset{+-}{\mu\nu};\rho}$ and $g_{\underset{+-}{\nu\mu};\rho}$. This leads to:

$$\frac{1}{2}\{(g_{\underset{+-}{\mu\nu};\rho} + g_{\underset{+-}{\rho\mu};\nu} + g_{\underset{+-}{\nu\rho};\mu}) - (g_{\underset{+-}{\nu\mu};\rho} + g_{\underset{+-}{\rho\nu};\mu} + g_{\underset{+-}{\mu\rho};\nu})\}$$

$$\equiv \{\partial_\mu \varphi_{\nu\rho} + \partial_\rho \varphi_{\mu\nu} + \partial_\nu \varphi_{\rho\mu}\} + 2\{\underset{\mathrm{v}}{\Gamma}_{\mu\nu,\rho} + \underset{\mathrm{v}}{\Gamma}_{\rho\mu,\nu} + \underset{\mathrm{v}}{\Gamma}_{\nu\rho,\mu}\} \tag{1.53}$$

where we have used (1.4) and defined*

$$\underset{\mathrm{v}}{\Gamma}_{\mu\nu,\rho} = \gamma_{\rho\sigma}\,\underset{\mathrm{v}}{\Gamma}{}^{\sigma}_{\mu\nu}\,. \tag{1.54}$$

The two identities (1.51) and (1.53) are completely analogous. One refers to the vector f^μ and the torsion Γ_ρ:

$$f^\mu = \frac{1}{\sqrt{-g}}\partial_\rho \mathfrak{g}^{\mu\rho}, \quad \Gamma_\rho = \underset{\mathrm{v}}{\Gamma}{}^{\sigma}_{\rho\sigma} \tag{1.55}$$

---

* We note here that the comma in $\Gamma_{\mu\nu,\rho}$ simply separates the symbols and is not a sign of differentiation.

the other to the pseudo-vector I^{μ} and the pseudo-torsion $\Gamma^*_\rho = \Gamma_{[\rho\sigma]}^{\sigma *}$ (with V under $\rho\sigma$):

$$I^{\mu} = \frac{1}{6}\,\epsilon^{\mu\nu\rho\sigma}(\partial_\nu \varphi_{\rho\sigma} + \partial_\sigma \varphi_{\nu\rho} + \partial_\rho \varphi_{\sigma\nu}) \qquad (1.56)$$

$$\Gamma^*_\rho = \underset{\mathrm{V}}{\Gamma}{}_{[\rho\sigma]}^{\sigma *} = \frac{\sqrt{-\gamma}}{2}\,\epsilon_{\rho\sigma\mu\nu}\,\gamma^{\mu\lambda}\,\gamma^{\nu\tau}\,\Gamma^{\sigma}_{\lambda\tau}$$

$$= -\frac{1}{2\sqrt{-\gamma}}\{\gamma_{\rho\mu}\,\gamma_{\sigma\nu}\,\epsilon^{\mu\nu\lambda\tau}\,\gamma^{\sigma\pi}\,\underset{\mathrm{V}}{\Gamma}{}_{\lambda\tau,\pi}\} \qquad (1.57)$$

$$= -\frac{1}{6\sqrt{-\gamma}}\,\gamma_{\rho\mu}\,\epsilon^{\mu\nu\lambda\tau}(\underset{\mathrm{V}}{\Gamma}{}_{\lambda\tau,\nu} + \underset{\mathrm{V}}{\Gamma}{}_{\nu\lambda,\tau} + \underset{\mathrm{V}}{\Gamma}{}_{\tau\nu,\lambda}).$$

C. THE CHOICE OF RICCI'S TENSOR IN A SPACE WITH A GENERAL AFFINE CONNECTION.

7. POSSIBLE CHOICES FOR THE CONTRACTED CURVATURE TENSOR.

In general Relativity, the contracted curvature tensor of order two can be defined without ambiguity. This is Ricci's tensor as given in (7).

In the case of a general affine connection, we must take into account the two possible forms of the parallel displacement of a vector:

$$\delta A^{\rho}_{+} = -\Gamma^{\rho}_{\mu\nu}\,A^{\mu}\,\delta x^{\nu} \qquad (1.36)$$

$$\delta A^{\rho}_{-} = -\Gamma^{\rho}_{\nu\mu}\,A^{\mu}\,\delta x^{\nu} = -\widetilde{\Gamma}^{\rho}_{\mu\nu}\,A^{\mu}\,\delta x^{\nu}. \qquad (1.37)$$

These two forms permit the definition of two tensors susceptible of generalizing the Riemann-Christoffel tensor (one being derived from $\Gamma^{\rho}_{\mu\nu}$ and the other from $\widetilde{\Gamma}^{\rho}_{\mu\nu}$:

$$R^{\rho}_{\mu\nu\sigma}(\Gamma) = \partial_\sigma\Gamma^{\rho}_{\mu\nu} - \partial_\nu\Gamma^{\rho}_{\mu\sigma} + \Gamma^{\lambda}_{\mu\nu}\,\Gamma^{\rho}_{\lambda\sigma} - \Gamma^{\lambda}_{\mu\sigma}\,\Gamma^{\rho}_{\lambda\nu} \qquad (1.58)$$

$$R^{\rho}_{\mu\nu\sigma}(\tilde{\Gamma}) = \partial_{\sigma}\tilde{\Gamma}^{\rho}_{\mu\nu} - \partial_{\nu}\tilde{\Gamma}^{\rho}_{\mu\sigma} + \tilde{\Gamma}^{\lambda}_{\mu\nu}\tilde{\Gamma}^{\rho}_{\lambda\sigma} - \tilde{\Gamma}^{\lambda}_{\mu\sigma}\tilde{\Gamma}^{\rho}_{\lambda\nu}$$

$$= \partial_{\sigma}\Gamma^{\rho}_{\nu\mu} - \partial_{\nu}\Gamma^{\rho}_{\sigma\mu} + \Gamma^{\lambda}_{\nu\mu}\Gamma^{\rho}_{\sigma\lambda} - \Gamma^{\lambda}_{\sigma\mu}\Gamma^{\rho}_{\nu\lambda}$$

(1. 59)

By contracting ρ and σ and ρ and μ, we obtain from (1. 58) two contracted tensors known as tensors of the first and second kind. A similar process applied to (1. 59) leads to two other tensors.

$$R_{\mu\nu} = R^{\rho}_{\mu\nu\rho}(\Gamma) = \partial_{\rho}\Gamma^{\rho}_{\mu\nu} - \partial_{\nu}\Gamma^{\rho}_{\mu\rho} + \Gamma^{\lambda}_{\mu\nu}\Gamma^{\rho}_{\lambda\rho} - \Gamma^{\lambda}_{\mu\rho}\Gamma^{\rho}_{\lambda\nu}, \qquad (1.60)$$

$$P_{\mu\nu} = R^{\rho}_{\rho\mu\nu}(\Gamma) = \partial_{\nu}\Gamma^{\rho}_{\rho\mu} - \partial_{\mu}\Gamma^{\rho}_{\rho\nu} \qquad (1.61)$$

$$\tilde{R}_{\mu\nu} = R^{\rho}_{\mu\nu\rho}(\tilde{\Gamma}) = \partial_{\rho}\Gamma^{\rho}_{\nu\mu} - \partial_{\nu}\Gamma^{\rho}_{\rho\mu} + \Gamma^{\lambda}_{\nu\mu}\Gamma^{\rho}_{\rho\lambda} - \Gamma^{\lambda}_{\rho\mu}\Gamma^{\rho}_{\nu\lambda} \qquad (1.62)$$

$$\tilde{P}_{\mu\nu} = R^{\rho}_{\rho\mu\nu}(\tilde{\Gamma}) = \partial_{\nu}\Gamma^{\rho}_{\mu\rho} - \partial_{\mu}\Gamma^{\rho}_{\nu\rho}\,. \qquad (1.63)$$

If the connection is symmetrical, it is clear that (1. 60) and (1. 62) coincide and that (1. 61) and (1. 63) are zero. For a general connection, this is not the case and, a priori, any combination of the above four tensors and the tensor

$$\Gamma_{\mu\nu} = \Gamma_{\mu}\Gamma_{\nu} \qquad (1.64)$$

can be considered.

8. CHOICE OF $R_{\mu\nu}$.

a. Hermitian principle.

To limit the arbitrariness in the choice of the tensor generalizing Ricci's tensor, Einstein first proposed the adoption of a hermitian principle (14). To this end, he defines the transposed quantities

$$\tilde{g}_{\mu\nu} = g_{\nu\mu} \tag{1.65}$$

$$\tilde{\Gamma}^{\rho}_{\mu\nu} = \Gamma^{\rho}_{\nu\mu} \tag{1.66}$$

According to Einstein, a tensor $A_{\mu\nu}(\Gamma, g)$ is said to be hermitian with respect to the indices μ and ν, if, upon interchanging μ and ν and changing g and Γ into $\tilde{g}$ and $\tilde{\Gamma}$ respectively, the new tensor is equal to the old one

$$A_{\mu\nu}(\Gamma, g) = A_{\nu\mu}(\tilde{\Gamma}, \tilde{g}). \tag{1.67}$$

The tensor is said to be antihermitian if

$$A_{\mu\nu}(\Gamma, g) = -A_{\nu\mu}(\tilde{\Gamma}, \tilde{g}). \tag{1.68}$$

According to this scheme, we can then reduce the basic tensors by considering only those linear combinations of (1.60), (1.61), (1.62) and (1.63) which are hermitian. We therefore would consider only the following tensors:

$$U_{\mu\nu} = \frac{1}{2}\{R_{\mu\nu} + \tilde{R}_{\nu\mu}\} - \frac{1}{4}\{P_{\mu\nu} + \tilde{P}_{\nu\mu}\} \tag{1.69}$$

$$H_{\mu\nu} = \frac{1}{2}\{P_{\mu\nu} + \tilde{P}_{\nu\mu}\} \tag{1.70}$$

$$\Gamma_{\mu\nu} = \Gamma_{\mu}\Gamma_{\nu}. \tag{1.64}$$

Explicitly:

$$U_{\mu\nu} = \partial_{\rho}\Gamma^{\rho}_{\mu\nu} - \frac{1}{2}\{\partial_{\mu}\Gamma^{\rho}_{\underline{\nu\rho}} + \partial_{\nu}\Gamma^{\rho}_{\underline{\mu\rho}}\} + \Gamma^{\lambda}_{\mu\nu}\Gamma^{\rho}_{\underline{\lambda\rho}} - \Gamma^{\lambda}_{\mu\rho}\Gamma^{\rho}_{\lambda\nu} \tag{1.71}$$

$$H_{\mu\nu} = \partial_{\mu}\Gamma_{\nu} - \partial_{\nu}\Gamma_{\mu}. \tag{1.72}$$

It seems somewhat advantageous to consider instead of $U_{\mu\nu}$ the combination

$$U_{\mu\nu} + \frac{1}{3} H_{\mu\nu} - \frac{1}{3} \Gamma_{\mu\nu}. \tag{1.73}$$

and we can see that this combination which, a priori, is really artificial, is exactly the tensor ${}^{(3)}R_{\mu\nu}$ given in Appendix II.

$${}^{(3)}R_{\mu\nu} = \frac{1}{2}\{ {}^{(2)}R_{\mu\nu} + {}^{(2)}\tilde{R}_{\nu\mu}\}, \tag{1.74}$$

${}^{(2)}R_{\mu\nu}$ being the Ricci tensor formed with the affine connection

$${}^{(2)}L^{\rho}_{\mu\nu} = \Gamma^{\rho}_{\mu\nu} + \frac{1}{3}\{\delta^{\rho}_{\mu}\ \Gamma_{\nu} - \delta^{\rho}_{\nu}\Gamma_{\mu}\} \tag{1.75}$$

whose torsion-vector is zero.

$${}^{(2)}L_{\rho} = {}^{(2)}L^{\sigma}_{\underset{\vee}{\rho\sigma}} = 0 \tag{1.76}$$

b. Connections with null torsions.

These results lead us to consider more or less complicated linear combinations of the Ricci tensor formed with the help of special affine connections, namely those with zero torsion vector. In fact, starting with the general connection $\Gamma^{\rho}_{\mu\nu}$, one can form the two connections

$${}^{(1)}L^{\rho}_{\mu\nu} = \Gamma^{\rho}_{\mu\nu} + \frac{2}{3}\delta^{\rho}_{\mu}\ \Gamma_{\nu} \tag{1.77}$$

$${}^{(2)}L^{\rho}_{\mu\nu} = \Gamma^{\rho}_{\mu\nu} + \frac{1}{3}\{\delta^{\rho}_{\mu}\ \Gamma_{\nu} - \delta^{\rho}_{\nu}\Gamma_{\mu}\}. \tag{1.78}$$

The anti-symmetrical parts of these connections are the same and in particular the torsion vector is null:

$$L_{\rho} = {}^{(1)}L^{\sigma}_{\underset{\vee}{\rho\sigma}} = {}^{(2)}L^{\sigma}_{\underset{\vee}{\rho\sigma}} = 0. \tag{1.79}$$

We can then define, from the affine connection $L^{\rho}_{\mu\nu}$, three tensors of the second order (54)

$${}^{(1)}R_{\mu\nu} = R^{\rho}_{\mu\nu\rho}({}^{1}L) \tag{1.80}$$

$$^{(2)}R_{\mu\nu} = R^{\rho}_{\mu\nu\rho}(^{2}L) \tag{1.81}$$

$$^{(3)}R_{\mu\nu} = \frac{1}{2}\{R^{\rho}_{\mu\nu\rho}(^{2}L) + R^{\rho}_{\nu\mu\rho}(^{2}\tilde{L})\} \tag{1.82}$$

of which only the third (1.82) is hermitian and is the tensor suggested by Einstein in the third edition (1950) of the Meaning of Relativity.

In short, if we do not retain the hermitian principle, it is possible to adopt a basic tensor which is some linear combination of eqs. (1.60) through (1.64) and, in particular one of the four tensors $R_{\mu\nu}$, $^{(1)}R_{\mu\nu}$, $^{(2)}R_{\mu\nu}$, $^{(3)}R_{\mu\nu}$. It is to the basic tensor that we will apply a variational principle:

In what follows, we shall use the tensor

$$R^{\rho}_{\mu\nu\rho} = \partial_{\rho}\Gamma^{\rho}_{\mu\nu} - \partial_{\nu}\Gamma^{\rho}_{\mu\rho} + \Gamma^{\lambda}_{\mu\nu}\Gamma^{\rho}_{\lambda\rho} - \Gamma^{\lambda}_{\mu\rho}\Gamma^{\rho}_{\lambda\nu}. \tag{1.60}$$

which we will call the Ricci tensor and to which we will apply the variational principle.

However, we shall examine in Appendix II the results of a variational principle applied to the tensors $^{(1)}R_{\mu\nu}$, $^{(2)}R_{\mu\nu}$ and $^{(3)}R_{\mu\nu}$ which are analogous to Ricci's tensor but which are formed with an affine connection whose torsion is null.

2

Field Equations
Variational Principles
Conservation Equations

A. METHOD OF APPLICATION OF A VARIATIONAL PRINCIPLE.

1. MIXED THEORY (EINSTEIN). VARIATION OF A FUNCTION ($\mathfrak{H}(g^{\mu\nu}, \Gamma^{\rho}_{\mu\nu}, \partial_\sigma \Gamma^{\rho}_{\mu\nu})$.

The mixed theory assumes that the variations of the affine connection and the metric of the space are independent of each other. We start from a variational principle applied to the scalar density $\mathfrak{H}$:

$$\delta \int \mathfrak{H} \, d\tau = 0 \tag{2.1}$$

$\mathfrak{H}$ being a function of the affine connection $\Gamma^{\rho}_{\mu\nu}$, and its first derivatives and the generalized field density $g^{\mu\nu}$. We assume that $\Gamma^{\rho}_{\mu\nu}$ and $g^{\mu\nu}$ are independent variables and that their variations $\delta\Gamma^{\rho}_{\mu\nu}$ and $\delta g^{\mu\nu}$ vanish at the limits of integration. With these assumptions, we obtain

$$\int \left[\frac{\partial \mathfrak{H}}{\partial \Gamma^{\rho}_{\mu\nu}} \delta\Gamma^{\rho}_{\mu\nu} + \frac{\partial \mathfrak{H}}{\partial (\partial_\sigma \Gamma^{\rho}_{\mu\nu})} \delta(\partial_\sigma \Gamma^{\rho}_{\mu\nu}) + \frac{\partial \mathfrak{H}}{\partial \mathcal{G}^{\mu\nu}} \delta\mathcal{G}^{\mu\nu} \right] d\tau = 0 \tag{2.2}$$

If we integrate the first two terms by parts, we have:

$$\int \left[\frac{\partial \mathfrak{H}}{\partial \Gamma^{\rho}_{\mu\nu}} \delta\Gamma^{\rho}_{\mu\nu} + \partial_{\sigma}\left(\frac{\partial \mathfrak{H}}{\partial(\partial_{\sigma}\Gamma^{\rho}_{\mu\nu})} \delta\Gamma^{\rho}_{\mu\nu} \right) - \left(\partial_{\sigma} \frac{\partial \mathfrak{H}}{\partial(\partial_{\sigma}\Gamma^{\rho}_{\mu\nu})} \right) \delta\Gamma^{\rho}_{\mu\nu} \right] d\tau . \qquad (2.3)$$

The second term of (2. 3) will not contribute to the integral since $\delta\Gamma_{\mu\nu}{}^{\rho}$ is zero at both limits. Thus (2. 1) can be written as

$$\int \{ G^{\mu\nu}_{\rho} \delta\Gamma^{\rho}_{\mu\nu} + J_{\mu\nu} \delta\mathcal{G}^{\mu\nu} \} d\tau = 0 \qquad (2.4)$$

where

$$G_{p}{}^{\mu\nu} = \frac{\partial H}{\partial \Gamma^{\rho}_{\mu\nu}} - \frac{\partial}{\partial x^{\sigma}} \left(\frac{\partial \mathfrak{H}}{\partial(\partial_{\sigma}\Gamma^{\rho}_{\mu\nu})} \right) \qquad (2.5)$$

$$J_{\mu\nu} = \frac{\partial \mathfrak{H}}{\partial \mathcal{G}^{\mu\nu}} . \qquad (2.6)$$

The basic equations of the theory will therefore be:

$$\begin{cases} \text{a.} \quad G^{\mu\nu}_{\rho} = 0 \\ \text{b.} \quad J_{\mu\nu} = 0. \end{cases} \qquad (2.7)$$

(2. 7a) are Euler's equations relative to the variation of the $\Gamma^{\rho}_{\mu\nu}$.

REMARK I. If we had imposed some conditions on the $\mathcal{G}^{\mu\nu}$, for example $\partial_{\rho} \mathcal{G}^{\mu\rho} = 0$, the variation $\delta\mathcal{G}^{\mu\nu}$ will not be independent and this would have to be taken into account in the variational principle. To this end, one can use the method of the Lagrange multipliers. If we impose the above condition to the $\mathcal{G}^{\mu\nu}$, we would have to add to H the scalar density $-2\sigma_{\mu} \partial_{\rho} \mathcal{G}^{\mu\rho}$, σ_{μ} being an arbitrary four vector. The variational principle leads to

$$\int \{ G^{\mu\nu}_{\rho} \delta\Gamma^{\rho}_{\mu\nu} + (J_{\mu\nu} + \partial_{\nu}\sigma_{\mu} - \partial_{\mu}\sigma_{\nu}) \delta\mathcal{G}^{\mu\nu} \} d\tau = 0 \qquad (2.8)$$

which leads to the field equations

$$\begin{cases} \text{a.} \quad G_{\rho}^{\mu\nu} = 0 \\ \text{b.} \quad \partial_{\rho}\, \mathfrak{J}^{\mu\nu} = 0 \\ \text{c.} \quad J_{\mu\nu} + \partial_{\nu}\sigma_{\mu} - \partial_{\mu}\sigma_{\nu} = 0. \end{cases} \tag{2.9}$$

REMARK II:

a. If we vary the scalar density

$$\mathfrak{H} = \mathcal{G}^{\mu\nu} R_{\mu\nu} \tag{2.10}$$

$R_{\mu\nu}$ being some tensor which is a function of the $\Gamma^{\rho}_{\mu\nu}$ and their derivatives, we would obtain, since $R_{\mu\nu} = \partial\, \mathfrak{H}/\partial\, \mathcal{G}^{\mu\nu} = J_{\mu\nu}$; for the field equations (2.7)

$$\begin{cases} \text{a.} \quad G_{\rho}^{\mu\nu} = 0 \\ \text{b.} \quad R_{\mu\nu} = 0 \end{cases} \tag{2.11}$$

b. In varying the scalar density

$$H' = H - 2\lambda\sqrt{-g}, \tag{2.12}$$

the variation of the $\mathcal{G}^{\mu\nu}$ gives the additional term

$$-2\lambda \frac{\partial\sqrt{-g}}{\partial \mathcal{G}^{\mu\nu}}\, \delta\mathcal{G}^{\mu\nu} = -\lambda\, \mathcal{G}_{\mu\nu}\, \delta\mathcal{G}^{\mu\nu}, \tag{2.13}$$

since

$$\frac{\partial\sqrt{-g}}{\partial \mathcal{G}^{\mu\nu}} = \frac{1}{2} g_{\mu\nu}.$$

The field equations will thus be:

$$\begin{cases} \text{a.} \quad G_{p}^{\mu\nu} = 0 \\ \text{b.} \quad R_{\mu\nu} - \lambda\, g_{\mu\nu} = 0. \end{cases} \tag{2.14}$$

2. PURE AFFINE THEORY (SCHRÖDINGER). VARIATION OF A FUNCTION $\mathfrak{H}(\Gamma^{\rho}_{\mu\nu}, \partial_{\sigma}\Gamma^{\rho}_{\mu\nu})$.

a. Instead of assuming that $\Gamma^{\rho}_{\mu\nu}$ and $\mathcal{G}^{\mu\nu}$ are independent variables, we can assume that the only independent variables are the $\Gamma^{\rho}_{\mu\nu}$. Thus $\mathfrak{H}$ depends only on $\Gamma^{\rho}_{\mu\nu}$ and their derivatives and for an increment $\delta\Gamma^{\rho}_{\mu\nu}$, the action integral leads to

$$\int G^{\mu\nu}_{\rho} \ \delta\Gamma^{\rho}_{\mu\nu} \ d\tau = 0. \tag{2.15}$$

Thus the only field equation derivable from the variational principle is equation (2.7a)

$$G^{\mu\nu}_{\rho} = 0. \tag{2.7a}$$

b. We can obtain a second group of equations by defining the $\mathcal{G}^{\mu\nu}$ from the scalar density $\mathfrak{H}$:

$$\mathcal{G}^{\mu\nu} = \frac{\partial \mathfrak{H}}{\partial R_{\mu\nu}}, \tag{2.16}$$

$R_{\mu\nu}$ being a function of $\Gamma^{\rho}_{\mu\nu}$ and their derivatives. It follows then that $\mathfrak{H}$ must be the homogeneous function

$$\mathfrak{H} = \mathcal{G}^{\mu\nu} R_{\mu\nu} \tag{2.17}$$

with the understanding that the variation $\delta\mathcal{G}^{\mu\nu}$ cannot yield an additional relation since the $\mathcal{G}^{\mu\nu}$ are not independent variables but uniquely defined by (2.16).

To proceed further, we must use (2.16) and consequently, we must give the explicit form of the action function in terms of $R_{\mu\nu}$. Most of the time, we adopt the function

$$\mathfrak{H} = \frac{2}{\lambda}\sqrt{-\det R_{\mu\nu}} \tag{2.18}$$

We can see then that (2.16) is equivalent to:

$$\mathcal{G}^{\mu\nu} = -\frac{1}{\lambda\sqrt{-\det R_{\mu\nu}}} \frac{\partial(\det R_{\mu\nu})}{\partial R_{\mu\nu}} \tag{2.19}$$

which leads to:

$$\frac{\text{minor of } g_{\mu\nu}}{\sqrt{-g}} = \frac{1}{\lambda} \frac{\text{minor of } R_{\mu\nu}}{\sqrt{-\det R_{\mu\nu}}}$$

$$R_{\mu\nu} = \lambda \, g_{\mu\nu}.$$

$$\begin{cases} \text{a.}\quad G_\rho^{\mu\nu} = 0 \\ \text{b.}\quad R_{\mu\nu} = \lambda\, g_{\mu\nu} \end{cases} \tag{2.20}$$

B. APPLICATION OF THE VARIATIONAL PRINCIPLE TO RICCI'S TENSOR $R_{\mu\nu}(\Gamma)$.

3. FIELD EQUATIONS DERIVED FROM A GENERAL AFFINE CONNECTION.

The strong system.

Starting with the principles of a mixed theory, we shall determine the variation of an assumed homogeneous scalar density.

$$\mathfrak{H}\,(\mathcal{G}^{\mu\nu}, \Gamma^{\rho}_{\mu\nu}, \partial_\sigma \Gamma^{\rho}_{\mu\nu}) = \mathcal{G}^{\mu\nu} R_{\mu\nu}. \tag{2.21}$$

We shall not impose a priori conditions to the variables. The field equations will then be given by (2.11)

$$\begin{cases} \text{a.}\quad G_\rho{}^{\mu\nu} = \dfrac{\partial \mathfrak{H}}{\partial \Gamma^{\rho}_{\mu\nu}} - \dfrac{\partial}{\partial x^\sigma} \dfrac{\partial \mathfrak{H}}{\partial(\partial_\sigma \Gamma^{\rho}_{\mu\nu})} = 0 \\ \text{b.}\quad R_{\mu\nu} = 0. \end{cases} \tag{2.11}$$

Let us write explicitly these equations in the case where $R_{\mu\nu}$ is Ricci's tensor

$$R_{\mu\nu} = \partial_\rho \Gamma^{\rho}_{\mu\nu} - \partial_\nu \Gamma^{\rho}_{\mu\rho} + \Gamma^{\sigma}_{\mu\nu} \Gamma^{\rho}_{\sigma\rho} - \Gamma^{\sigma}_{\mu\rho} \Gamma^{\rho}_{\sigma\nu}.$$

(2.11a) then becomes:*

$$G_\alpha^{\beta\gamma} \equiv \mathcal{G}^{\mu\nu} \frac{\partial R_{\mu\nu}}{\partial \Gamma^{\alpha}_{\beta\gamma}} - \frac{\partial}{\partial x^\epsilon} \left[g^{\mu\nu} \frac{\partial R_{\mu\nu}}{\partial(\partial_\epsilon \Gamma^{\alpha}_{\beta\gamma})} \right] = 0 \tag{2.22}$$

---

* We can also generalize Palatini's method for a nonsymmetrical connection. We then have

$$\delta R_{\mu\nu} = \left(\delta \Gamma^{\;\;\overset{0}{\rho}}_{\underset{+}{\mu}\underset{-}{\nu}}\right)_{;\rho} - \frac{1}{2}\left(\delta \Gamma^{\rho}_{\underset{\mp}{\underline{\mu\rho}}}\right)_{;\nu} - \frac{1}{2}\left(\delta \Gamma^{\rho}_{\underset{-}{\underline{\nu\rho}}}\right)_{;\mu}.$$

See Einstein (14) p. 143.

with

$$\frac{\partial R_{\mu\nu}}{\partial(\partial_\epsilon \Gamma^\alpha_{\beta\gamma})} = \delta^\epsilon_\rho \; \delta^\beta_\mu \; \delta^\gamma_\nu \; \delta^\rho_\alpha - \delta^\epsilon_\nu \; \delta^\beta_\mu \; \delta^\gamma_\rho \; \delta^\rho_\alpha , \tag{2.23}$$

$$\frac{\partial R_{\mu\nu}}{\partial \Gamma^\alpha_{\beta\gamma}} = \delta^\rho_\mu \; \delta^\gamma_\nu \; \delta^\sigma_\alpha \; \Gamma^\rho_{\sigma\rho} + \delta^\beta_\sigma \; \delta^\gamma_\rho \; \delta^\rho_\alpha \; \Gamma^\sigma_{\mu\nu}$$

$$- \delta^\beta_\mu \; \delta^\gamma_\rho \; \delta^\sigma_\alpha \; \Gamma^\rho_{\sigma\nu} - \delta^\beta_\sigma \; \delta^\gamma_\nu \; \delta^\rho_\alpha \; \Gamma^\sigma_{\mu\rho} \tag{2.24}$$

whence

$$- G^{\beta\gamma}_\alpha = \partial_\alpha \mathcal{G}^{\beta\gamma} - \delta^\gamma_\alpha \; \partial_\nu \mathcal{G}^{\beta\nu} - \mathcal{G}^{\beta\gamma} \Gamma^\rho_{\alpha\rho} - \delta^\gamma_\alpha \; \mathcal{G}^{\mu\nu} \Gamma^\beta_{\mu\nu}$$

$$+ \mathcal{G}^{\beta\gamma} \; \Gamma^\gamma_{\alpha\nu} + \mathcal{G}^{\mu\gamma} \; \Gamma^\beta_{\mu\alpha} = 0 \tag{2.25}$$

contracting α and γ and then α and β leads to:

$$- 3 \; \partial_\alpha \mathcal{G}^{\beta\alpha} - \beta \; \mathcal{G}^{\beta\alpha} \; \Gamma^\rho_{\underset{\text{V}}{\alpha\rho}} - 3 \mathcal{G}^{\mu\nu} \; \Gamma^\beta_{\mu\nu} = 0 \tag{2.26}$$

$$\partial_\alpha \; \mathfrak{g}^{\alpha\gamma} = 0, \quad (\mathcal{G}^{\alpha\beta} = \mathcal{H}^{\alpha\beta} + \mathfrak{g}^{\alpha\beta}). \tag{2.27}$$

We solve (2.26) for $\mathcal{G}^{\mu\nu} \; \Gamma_{\mu\nu}{}^\beta$ and substitute its value in (2.25), obtaining:

$$\partial_\alpha \; \mathcal{G}^{\beta\gamma} - \mathcal{G}^{\beta\gamma} \; \Gamma^\rho_{\alpha\rho} + \frac{2}{3} \; \delta^\gamma_\alpha \; \mathcal{G}^{\beta\sigma} \; \Gamma_\sigma + \mathcal{G}^{\beta\sigma} \; \Gamma^\gamma_{\alpha\sigma}$$

$$+ \mathcal{G}^{\sigma\gamma} \; \Gamma^\beta_{\sigma\alpha} = 0 \tag{2.28}$$

whence

$$\mathcal{G}^{\underset{+-}{\beta\gamma}} ; \alpha + \frac{2}{3} \; \delta^\gamma_\alpha \; \mathcal{G}^{\beta\sigma} \; \Gamma_\sigma - \mathcal{G}^{\beta\gamma} \; \Gamma_\alpha = 0. \tag{2.29}$$

The field equations in a space with a general affine connection will then be

$$\text{a.} \quad \mathcal{G}^{\underset{+-}{\mu\nu}} ;\rho = - \frac{2}{3} \; \delta^\nu_\rho \; \mathcal{G}^{\mu\sigma} \; \Gamma_\sigma + \mathcal{G}^{\mu\nu} \; \Gamma_\rho, \quad \partial_\mu \; \mathfrak{g}^{\mu\rho} = 0$$

$$\text{b.} \quad R_{\mu\nu} = 0. \tag{I}$$

Equations (I) are deduced from a variational principle and are necessarily compatible. Einstein suggested that one should add to I, the condition

$$\Gamma_\rho = 0.$$

This leads to $\mathcal{G}^{\underset{+\,-}{\mu\nu}}{}_{;\rho} = 0$ and as a consequence of this condition (1. 51) leads to $\partial_\mu \mathfrak{I}^{\mu\rho} = 0$. Thus we have the following system:

$$\begin{aligned} &\mathcal{G}^{\underset{+\,-}{\mu\nu}}{}_{;\rho} = 0 \qquad \Gamma_\rho = 0 \;(\rightarrow \partial_\mu \mathfrak{I}^{\mu\rho} = 0) \\ &R_{\mu\nu} = 0 \end{aligned} \tag{F}$$

These are the equations of a strong system and are not deducible from a variational principle.

4. CHANGE OF AFFINE CONNECTION.

The weak system.

We shall define a new affine connection $L^\rho_{\mu\nu}$ such that $L_\rho = L^\sigma_{\underset{\vee}{\rho\sigma}} = 0$ and refer all the field equations to it. We shall adopt the following notations:

			$L^\rho_{\mu\nu}$	$\Delta^\rho_{\mu\nu}$
Affine connection	$\left\{{\rho \atop \mu\nu}\right\}$	$\Gamma^\rho_{\mu\nu}$	$(L_\rho = 0)$	$(g_{\underset{+\,-}{\mu\nu};\rho} = 0)$
Covariant differentiation	∇_ρ	;	D_ρ	$;\rho$
Ricci's tensor	$G_{\mu\nu}$	$R_{\mu\nu}$	$W_{\mu\nu}$	
Einstein's tensor	$S_{\mu\nu}$	$H_{\mu\nu}$	$K_{\mu\nu}$	

(2.30)

Let

$$L^\rho_{\mu\nu} = \Gamma^\rho_{\mu\nu} + \frac{2}{3}\,\delta^\rho_\mu\,\Gamma_\nu, \tag{2. 31}$$

such that

$$L_\mu = L^\rho_{\underset{\vee}{\mu\rho}} = 0. \tag{2. 32}$$

Equations (2. 28) can then be written as

$$\partial_\alpha \mathcal{G}^{\beta\gamma} - \mathcal{G}^{\beta\gamma} L^{\rho}_{\alpha\rho} + \mathcal{G}^{\beta\sigma} L^{\gamma}_{\alpha\sigma} + \mathcal{G}^{\sigma\gamma} L^{\beta}_{\sigma\alpha} = 0 \qquad (2.33)$$

or

$$D_\alpha \mathcal{G}^{\overset{\beta\gamma}{+-}} = 0 \qquad (2.34)$$

$R_{\mu\nu}$ as given in (1.60) takes the form:

$$R_{\mu\nu} = W_{\mu\nu} - \frac{2}{3}\{\partial_\mu \Gamma_\nu - \partial_\nu \Gamma_\mu\} = 0 \quad (2.35)$$

where $W_{\mu\nu}$ is the Ricci tensor formed with $L^{\rho}_{\mu\nu}$:

$$W_{\mu\nu} = \partial_\rho L^{\rho}_{\mu\nu} - \partial_\nu L^{\rho}_{\mu\rho} + L^{\lambda}_{\mu\nu} L^{\rho}_{\lambda\rho} - L^{\lambda}_{\mu\rho} L^{\rho}_{\lambda\nu}. \quad (2.36)$$

We can, from (2. 34), determine all the $L^{\rho}_{\mu\nu}$. But (2. 31) shows that $\Gamma^{\rho}_{\mu\nu}$ can be defined up to some arbitrary Γ_ρ; thus the antisymmetrical part of (2. 35) must satisfy the relations

$$\partial_\rho W_{\underset{\vee}{\mu\nu}} + \partial_\nu \Gamma_{\underset{\vee}{\rho\mu}} + \partial_\mu W_{\underset{\vee}{\nu\rho}} = 0. \qquad (2.37)$$

Finally, the field equations can be obtained from I* with the understanding that the affine connection in $L^{\rho}_{\mu\nu}$ ($L_\rho = L_{\underset{\vee}{\rho\sigma}}{}^{\sigma} = 0$):

$$\text{a.} \quad D_\rho \mathcal{G}^{\overset{\mu\nu}{+-}} = 0 \qquad \partial_\rho \mathfrak{g}^{\mu\rho} = 0$$

$$\text{b.} \quad W_{\underline{\mu\nu}} = 0 \qquad \partial_\rho W_{\underset{\vee}{\mu\nu}} + \partial_\nu W_{\underset{\vee}{\rho\mu}} + \partial_\mu W_{\underset{\vee}{\nu\rho}} = 0 \qquad (II)$$

---

* The equations equivalent to (I) but referred to $L^{\rho}_{\mu\nu}$ are (IIa) and (2. 35)

$$D_\rho g^{\overset{\mu\nu}{+-}} = 0 \qquad \partial_\rho \mathfrak{g}^{\mu\rho} = 0$$

$$(I') \qquad W_{\mu\nu} = \frac{2}{3}(\partial_\mu \Gamma_\nu - \partial_\nu \Gamma_\mu)$$

They identify the arbitrary vector for which $W_{\mu\nu}$ is a rotational tensor with the torsion (10) page 267, (43).

There are the equations for a 'weak' system. They are completely derivable from a variational principle.*

If we add a cosmological term to the action function or if, in a purely affine theory, we choose the density (2.18), we would obtain equations of type II on the condition of replacing $W_{\mu\nu}$ by the tensor

$$\overline{W}_{\mu\nu} = W_{\mu\nu} - \lambda\, g_{\mu\nu}. \tag{2.38}$$

We would then have

$$\begin{cases} \text{a.} \quad D_\rho \overset{\mu\nu}{\mathcal{G}}{}_{+-} = 0 \qquad \partial_\rho \mathfrak{J}^{\mu\rho} = 0 \\ \text{b.} \quad \underline{W_{\mu\nu}} = \lambda\, \gamma_{\mu\nu} \qquad \partial_\rho \underset{\vee}{W_{\mu\nu}} + \partial_\nu \underset{\vee}{W_{\rho\mu}} + \partial_\mu \underset{\vee}{W_{\nu\rho}} = \lambda\, \varphi_{\mu\nu\rho} \end{cases} \tag{II'}$$

This is the system of equations adopted by Schrödinger.

5. REMARKS ON THE SYSTEMS OF EQUATIONS DEDUCIBLE FROM THE VARIATION OF $\mathfrak{H} = \mathcal{G}^{\mu\nu}\, {}^{(a)}R_{\mu\nu}$.

In Appendix II, we will apply the variational principle to a Ricci tensor ${}^{(a)}R_{\mu\nu}(a = 1,\ 2,\ 3)$ formed with the affine connection L whose torsion vector is zero (54). We obtain then a system of equations which are weaker than (II). In fact, a transformation of the affine connection permits us to express them in terms of a connection Δ (See (II - 7) App. II) and to give them the form

$$\begin{cases} \text{a.} \quad \overset{\mu\nu}{\mathcal{G}}{}_{+-;\rho} = 0 \\ \text{b.} \quad R_{\mu\nu}({}^{(a)}\Delta) + {}^{(a)}K_{\mu\nu}(f^\rho) = 0 \end{cases} \tag{III}$$

* Mrs. J. Winogradski has shown that all hamiltonians of the type $\mathcal{G}^{\mu\nu} R_{\mu\nu}$ are equivalent for the formulation of the field equations. In fact the densities $\mathcal{G}^{\mu\nu} R_{\mu\nu}$, $\mathcal{G}^{\mu\nu} R_{\nu\mu}$, $\mathcal{G}^{\mu\nu} \tilde{R}_{\mu\nu}$ and $\mathcal{G}^{\mu\nu} \tilde{R}_{\nu\mu}$ can be deduced from each other by a λ or $\tilde{\lambda}$ transformation, i. e., in substituting for $\Gamma^\rho_{\mu\nu}$ the connection $\Gamma^\rho_{\mu\nu} + \delta^\rho_\mu\ \lambda_\nu$ (or $\Gamma^\rho_{\mu\nu} + \delta^\rho_\nu\ \Gamma_\mu$). The field equations obtained by varying the metric under the condition that $\partial_\rho \mathfrak{J}^{\mu\rho} = 0$ and the connection are invariant under these λ-transformations. The transformation of the affine connection (2.31) which leads to the field equations (II) is a particular case of the λ transformations (cf. (43) and Einstein (19)).

We can make two remarks with respect to these equations.

I - From the variational principle, it is clear that $\Gamma^{\rho}_{\mu\nu}$ can be determined up to an arbitrary vector Γ_{ρ}.*

II - If we define a connection $\Delta^{\rho}_{\mu\nu}$ such that the covariant derivative of $\mathcal{G}^{\mu\nu}$ is zero with respect to this connection, we have the following results:

a. From the identities (1. 51), we have:

$$\partial_{\rho}\, \mathcal{J}^{\mu\rho} - \mathcal{H}^{\mu\rho}\, \Delta_{\rho} = 0 \tag{2. 39}$$

If we multiply by $1/\sqrt{-g}\ h_{\mu\sigma}$ and sum, we obtain

$$\Delta_{\sigma} = \frac{h_{\mu\sigma}}{\sqrt{-g}}\ \partial_{\rho}\, (\sqrt{-g}\, f^{\mu\rho}) = h_{\mu\sigma}\, f^{\mu}. \tag{2. 40}$$

We can substitute for $h_{\mu\sigma}$ its value as given in (1. 20s) and we are led to

$$\Delta_{\sigma} = (\gamma_{\mu\sigma} + \varphi_{\mu\lambda}\, \varphi_{\sigma\tau}\, \gamma^{\lambda\tau})\, f^{\mu} = f_{\sigma} - f_{\overset{=}{\sigma}} \tag{2. 41}$$

where $f_{\overset{=}{\sigma}} = \varphi_{\sigma\tau}\gamma^{\tau\lambda}\varphi_{\lambda\mu} f^{\mu}$

b. From the identities (1. 53), we have

$$\Delta_{\mu\nu\rho} = -\frac{1}{2}\, \varphi_{\mu\nu\rho} \tag{2. 42}$$

with

$$\Delta_{\mu\nu\rho} = \underset{\vee}{\Delta}_{\mu\nu,\,\rho} + \underset{\vee}{\Delta}_{\rho\mu,\,\nu} + \underset{\vee}{\Delta}_{\nu\rho,\,\mu} \tag{2. 43}$$

$$\varphi_{\mu\nu\rho} = \partial_{\rho}\, \varphi_{\mu\nu} + \partial_{\nu}\, \varphi_{\rho\mu} + \partial_{\mu}\, \varphi_{\nu\rho}. \tag{2. 44}$$

It is clear that the 64 coefficients of the affine connection are determined unequivocally from (IIIa).

---

* This is due to the fact that $\Gamma^{\rho}_{\mu\nu}$ never satisfies the equations $\mathcal{G}^{\underset{+}{\mu}\underset{-}{\nu}}{}_{;\rho} = 0$, which would then determine it completely, unless additional conditions are imposed on it.

SPECIAL CASE: In the special case where we start from the Ricci tensor formed with a generalized affine connection, the connection Δ such that $D_\rho \mathcal{G}^{\mu\nu}_{+-} = 0$ is the same as the connection $^{(1)}L^\rho_{\mu\nu} = \Gamma^\rho_{\mu\nu} + 2/3\ \delta^\rho_\mu \Gamma_\nu$ whose torsion is zero. In this case, it turns out that the four vector $\mathcal{J}^\mu$ is zero. (1.51) written in terms of $L^\rho_{\mu\nu}$ is automatically satisfied by the simultaneous vanishing of L_ρ and $\mathcal{J}^\rho$. The connection L_ρ which satisfies equations (II) is also determined unequivocally. The vanishing of its torsion vector results directly from these equations.

In all our further developments, we shall use Ricci's tensor as the basic tensor. The condition $\mathcal{J}^\rho = 0$ results directly from the variational principle and the connection $\Delta^\rho_{\mu\nu}$ coincides with the connection $L^\rho_{\mu\nu}$ whose torsion vector is zero. The field equations which result from a variational principle and which we shall use are given by (II):

$$\begin{cases} \text{a.}\quad D_\rho\, \mathcal{G}^{\mu\nu}_{+-} = 0 \qquad L_\rho = \underset{\vee}{L_{\rho\sigma}} = 0 \sim \mathcal{J}^\rho = 0 \\ \text{b.}\quad W_{\underline{\mu\nu}} = 0 \qquad \partial_\rho \underset{\vee}{W_{\mu\nu}} + \partial_\nu \underset{\vee}{W_{\rho\mu}} + \partial_\mu \underset{\vee}{W_{\nu\rho}} = 0. \end{cases} \tag{II}$$

It will be in this context that we shall treat most of the applications and that we shall discuss the possible forms of the spherically symmetric solutions.*

c. Conservation equations (41), (115), (116).

In a Riemannian space, the tensor S^ν_μ (which is a function of the $g_{\mu\nu}$, and its first and second derivatives) satisfies the identity

$$\nabla_\rho S^\rho_\mu = 0 \tag{2.45}$$

* Nevertheless, we must not forget the possibilities resulting from the use of a different basic tensor (such as $^{(1)}R$) than the Ricci tensor. The field equations will then be given by eq. (IIIa) with $\Delta_\rho = f_\rho - f_{\bar{\rho}}$. The disappearance of the conditions $f_\rho = 0$ or $\Delta_\rho = 0$ might entail useful modifications for the determination of solution.

and will necessarily have the form

$$S_{\mu}^{\rho} = G_{\mu}^{\rho} - \frac{1}{2}\delta_{\mu}^{\rho}(G - 2\lambda). \qquad (2.46)$$

If $\lambda = 0$, (2.45) takes the form

$$\partial_{\rho}\mathfrak{S}_{\mu}^{\rho} + \frac{1}{2}\mathcal{G}_{\rho\sigma}\partial_{\mu}\mathcal{G}^{\rho\sigma} = 0 \qquad (2.47)$$

These identities follow uniquely from the definition of Ricci's tensor as a function of the $g_{\mu\nu}$ and their derivatives, i.e., from the definition of the Christoffel symbols equivalent to $\Delta_{\rho}\mathcal{G}^{\mu\nu} = 0$.

In unified theory, the definition of the affine connection in terms of the metric is equivalent to equations

$$D_{\rho}\mathcal{G}^{\underset{+-}{\mu\nu}} = 0 \qquad \partial_{\rho}\mathcal{g}^{\mu\rho} = 0. \qquad \text{(IIa)}$$

The question arises as to whether Eq. IIa is sufficient to lead to identities between the components of Ricci's tensor corresponding to the connection $\Gamma_{\mu\nu}^{\rho}$.

We will give here the proof of Lichnerowicz (41). Consider the integral

$$\mathcal{A} = \int R_{\mu\nu}\mathcal{G}^{\mu\nu}\,d\tau. \qquad (2.48)$$

Generalizing Weyl's method, we determine the variation of $\mathcal{G}^{\mu\nu}$ resulting from an infinitesimal change of coordinates $x'^{\rho} = x^{\rho} + \xi^{\rho}$ determined by the vector field ξ^{ρ}. We will have:

$$\delta g^{\mu\nu} = -(\partial_{\rho}g^{\mu\nu})\xi^{\rho} + g^{\mu\rho}\partial_{\rho}\xi^{\nu} + g^{\rho\nu}\partial_{\rho}\xi^{\mu}.^{(1)} \qquad (2.49)$$

These equations determine the corresponding variations:

$$-\delta\sqrt{-g} = \frac{\sqrt{-g}}{2}g_{\mu\nu}\,\delta g^{\mu\nu} = (\partial_{\rho}\sqrt{-g})\xi^{\rho} + \sqrt{-g}\,\partial_{\rho}\xi^{\rho}$$

$$= \partial_{\rho}(\sqrt{-g}\,\xi^{\rho}), \qquad (2.50)$$

$$-\delta(R_{\mu\nu}\mathcal{G}^{\mu\nu}d\tau) = -\{\sqrt{-g}\ \delta R + R\delta\sqrt{-g}\}d\tau$$

$$= \{\sqrt{-g}\,(\partial_\rho R)\xi^\rho + R\partial_\rho(\sqrt{-g}\,\xi^\rho)\}\,d\tau$$

$$= \partial_\rho(R\sqrt{-g}\xi^\rho)\,d\tau. \tag{2.51}$$

The variation $\delta\mathcal{A}$ is expressed in terms of the integral of a divergence. From Stokes' formula, this integral is zero unless the increments ξ^ρ are zero at the limits of integration. We assume that the latter holds and obtain

$$\delta\mathcal{A} = \int\{(\delta R_{\mu\nu})\,\mathcal{G}^{\mu\nu} + R_{\mu\nu}\,\delta\mathcal{G}^{\mu\nu}\}d\tau = 0. \tag{2.52}$$

However, Eqs. (IIa) which we assume to hold are precisely the necessary and sufficient condition for

$$\int(\delta R_{\mu\nu})\,\mathcal{G}^{\mu\nu}\,d\tau = 0 \tag{2.53}$$

for an increment $\delta\Gamma^\rho_{\sigma\tau}$ which vanish at the limits of integration. This will be the case if we assume that the ξ^ρ and their first two derivatives also vanish at the limits. We will then have:

$$\int R_{\mu\nu}\delta\ \mathcal{G}^{\mu\nu}\,d\tau = 0 \tag{2.54}$$

---

* $\delta g^{\mu\nu}$ is then determined by the difference between the $g^{\mu\nu}$ computed at the two points which have the coordinates x^ρ in the two reference systems.

$$\delta g^{\mu\nu} = g'^{\mu\nu}(x) - g^{\mu\nu}(x) = [g'^{\mu\nu}(x') - g^{\mu\nu}(x)] - [g'^{\mu\nu}(x') - g'^{\mu\nu}(x)]$$

$$= \left[\frac{\partial x'^\mu}{\partial x^\rho}\frac{\partial x'^\nu}{\partial x^\sigma}\,g^{\rho\sigma}(x) - g^{\mu\nu}(x)\right] - \frac{\partial g^{\mu\nu}}{\partial x^\rho}\,\xi^\rho$$

$$= \left[\left(\delta_\rho{}^\mu + \frac{\partial\xi^\mu}{\partial x^\rho}\right)\left(\delta_\sigma{}^\nu + \frac{\partial\xi^\nu}{\partial x^\sigma}\right)g^{\rho\sigma} - g^{\mu\nu}(x)\right] - \frac{\partial g^{\mu\nu}}{\partial x^\rho}\,\xi^\rho$$

$$= g^{\mu\sigma}\frac{\partial\xi^\nu}{\partial x^\sigma} + g^{\sigma\nu}\frac{\partial\xi^\mu}{\partial x^\sigma} - \frac{\partial g^{\mu\nu}}{\partial x^\rho}\,\xi^\rho.$$

For a scalar R:

$$\delta R = R'(x) - R(x) = -[R'(x') - R'(x)] = -\frac{\partial R}{\partial x^\rho}\,\xi^\rho.$$

(cf. Von Laue (9), p. 208).

But, from (2. 49):

$$-R_{\mu\nu}\ \delta\mathcal{G}^{\mu\nu} = R_{\mu\nu}(\partial_\rho \mathcal{G}^{\mu\nu})\,\xi^\rho - (R_{\mu\nu}\mathcal{G}^{\rho\nu} + R_{\nu\mu}\mathcal{G}^{\nu\rho})\,\partial_\rho\xi^\mu + R_{\mu\nu}\ \mathcal{G}^{\mu\nu}\,\partial_\rho\,\xi^\rho$$

$$= \partial_\rho\,[R_{\mu\nu}\mathcal{G}^{\mu\nu}\xi^\rho - (R_{\mu\nu}\mathcal{G}^{\rho\nu} + R_{\nu\mu}\mathcal{G}^{\nu\rho})\,\xi^\mu]$$

$$+[\partial_\rho\,(R_{\mu\nu}\mathcal{G}^{\rho\nu} + R_{\nu\mu}\mathcal{G}^{\nu\rho}) - \delta_\mu^\rho\,\mathcal{G}^{\alpha\beta}\partial_\rho R_{\alpha\beta}\,]\xi^\mu.$$

The integral of this expression must be zero independently of ξ^μ. The first term being a divergence will vanish. We will thus have:

$$\partial_\rho(R_{\mu\nu}\mathcal{G}^{\rho\nu} + R_{\nu\mu}\mathcal{G}^{\nu\rho}) - \delta_\mu^\rho\ \mathcal{G}^{\alpha\beta}\partial_\rho\,R_{\alpha\beta} = 0. \quad (2.\ 55)$$

Let

$$\mathcal{H}_\mu^\rho = \sqrt{-g}\ H_\mu^\rho\ , \qquad H_\mu^\rho = \frac{1}{2}\,(R_{\mu\nu}g^{\rho\nu} + R_{\nu\mu}g^{\nu\rho}) - \frac{1}{2}\,\delta_\mu^\rho R. \quad (2.\ 56)$$

We then have:*

$$\partial_\rho\mathcal{H}_\mu^\rho + \frac{1}{2}\,R_{\alpha\beta}\partial_\mu\mathcal{G}^{\alpha\beta} = 0. \quad (2.\ 57)$$

This is an identity of the type (2. 47). It refers to the tensor $R_{\mu\nu} = R_{\mu\nu}(\Gamma)$.

To write it in terms of the tensor $W_{\mu\nu} = R_{\mu\nu}(L)$, we make the substitution

$$R_{\mu\nu} = W_{\mu\nu} - \frac{2}{3}\,(\partial_\mu\Gamma_\nu - \partial_\nu\Gamma_\mu). \quad (2.\ 35)$$

We will then have:

$$\mathcal{H}_\mu^\rho = \mathcal{K}_\mu^\rho - \frac{2}{3}\{\Psi_{\mu\nu}\ \mathcal{g}^{\rho\nu} - \frac{1}{2}\ \delta_\mu^\rho\ \Psi_{\alpha\beta}\ \mathcal{g}^{\alpha\beta}\}, \quad (2.\ 58)$$

by letting

$$\Psi_{\mu\nu} = \partial_\mu\Gamma_\nu - \partial_\nu\Gamma_\mu, \quad (2.\ 59)$$

---

* cf. Schrodinger, Proc. Roy. Ir. Acad. 51A, 6, 1948 Eq. (4.1); Lichnerowicz, C. R., Acad. Sc. 237, 1383, 1953, Eq. (4).

$$\mathcal{K}_{\mu}^{\rho} = \sqrt{-g}\ K_{\mu}^{\rho}, \quad K_{\mu}^{\rho} = \frac{1}{2}\{W_{\mu\nu}\, g^{\rho\nu} + W_{\nu\mu} g^{\nu\rho}\} - \frac{1}{2}\,\delta_{\mu}^{\rho}\, W. \tag{2.60}$$

Whence, substituting (2. 35) and (2. 58) in (2. 57) and recalling that $\partial_{\rho}\, \mathcal{G}^{\mu\rho} = 0$,

$$\partial_{\rho}\, \mathcal{K}_{\mu}^{\rho} + \frac{1}{2} W_{\alpha\beta}\, \partial_{\mu}\, \mathcal{G}^{\alpha\beta} = 0. \tag{2.61}$$

We will thus have identities of the same form in the case of Ricci's tensor expressed in terms of the $L_{\mu\nu}^{\rho}$.

3

The First Group of Einstein's Equations

EXPRESSION OF THE AFFINE CONNECTION AS A FUNCTION OF THE FIELDS.
(53), (54), (55), (57).

The group (IIa) or (IIIa) of Einstein's equations permit the determination of the affine connection in terms of the fields $g_{\mu\nu}$.

We shall solve the system

$$\mathcal{G}^{\mu\nu}_{+-;\rho} = 0 \tag{3. 1}$$

without making any assumptions about the affine connection $\Delta^{\rho}_{\mu\nu}$ in terms of which (3. 1) is expressed. We shall determine* the most general affine connection satisfying (3. 1).

The solution of the system (IIa)

$$D_{\rho}\, \mathcal{G}^{\mu\nu}_{+-} = 0, \qquad \partial_{\rho}\, \mathfrak{g}^{\mu\rho} = 0, \qquad L_{\rho} = 0 \tag{IIa}$$

follows immediately from those of (3. 1) by setting the four-vector f_{ρ} equal to zero.

---

* This is the solution of $\mathcal{G}^{\mu\nu}_{+-;\rho} = 0$ which we labeled $\Gamma^{\rho}_{\mu\nu}$, if no confusion arises when a single affine connection is introduced. (J. Phys. Rad., 13, 177, 1952; 16, 21, 1955.)

NOTATIONS: Consider a four-vector with components A^σ. We will use the following notation:

$$\left\{\begin{aligned}
A_\rho &= \gamma_{\rho\sigma} A^\sigma, \\
A_{\bar{\rho}} &= \varphi_{\rho\sigma} A^\sigma = \varphi_{\rho\sigma}\gamma^{\sigma\tau}A_\tau, \\
A_{\bar{\bar{\rho}}} &= \varphi_{\rho\sigma} A^{\bar{\sigma}} = \varphi_{\rho\sigma}\gamma^{\sigma\tau}A_{\bar{\tau}} = \varphi_{\rho\sigma}\gamma^{\sigma\tau}\varphi_{\tau\lambda}\gamma^{\lambda\mu}A_\mu, \\
A_{\bar{\bar{\bar{\rho}}}} &= \varphi_{\rho\sigma} A^{\bar{\bar{\sigma}}} = \varphi_{\rho\sigma}\gamma^{\sigma\tau}\varphi_{\tau\lambda}\gamma^{\lambda\mu}\varphi_{\mu\pi}\gamma^{\pi\delta}A_\delta, \\
A_{\bar{\bar{\bar{\bar{\rho}}}}} &= \varphi_{\rho\sigma} A^{\bar{\bar{\bar{\sigma}}}} = \varphi_{\rho\sigma}\gamma^{\sigma\tau}\varphi_{\tau\lambda}\gamma^{\lambda\mu}\varphi_{\mu\pi}\gamma^{\pi\delta}\varphi_{\delta\eta}\gamma^{\eta\nu}A_\nu, \\
&\dots
\end{aligned}\right. \tag{3.2}$$

and

$$\left\{\begin{aligned}
A^\rho &= \gamma^{\rho\sigma} A_\sigma, \\
A^{\bar{\rho}} &= \gamma^{\rho\sigma} A_{\bar{\sigma}} = \gamma^{\rho\sigma}\varphi_{\sigma\tau}A^\tau, \\
A^{\bar{\bar{\rho}}} &= \gamma^{\rho\sigma} A_{\bar{\bar{\sigma}}} = \gamma^{\rho\sigma}\varphi_{\sigma\tau}\gamma^{\tau\lambda}\varphi_{\rho\mu}A^\mu, \\
A^{\bar{\bar{\bar{\rho}}}} &= \gamma^{\rho\sigma} A_{\bar{\bar{\bar{\sigma}}}} = \gamma^{\rho\sigma}\varphi_{\sigma\tau}\gamma^{\tau\lambda}\varphi_{\lambda\mu}\gamma^{\mu\pi}\varphi_{\pi\delta}A^\delta, \\
A^{\bar{\bar{\bar{\bar{\rho}}}}} &= \gamma^{\rho\sigma} A_{\bar{\bar{\bar{\bar{\sigma}}}}} = \gamma^{\rho\sigma}\varphi_{\sigma\tau}\gamma^{\tau\lambda}\varphi_{\lambda\mu}\gamma^{\mu\pi}\varphi_{\pi\delta}\gamma^{\delta\eta}\varphi_{\eta\nu}A^\nu, \\
&\dots
\end{aligned}\right. \tag{3.3}$$

One can establish the following identities between these expressions (see App. III)

$$A_{\bar{\bar{\bar{\rho}}}} = -\frac{\varphi}{\gamma}A_\rho - \frac{1}{\gamma}(g - \gamma - \varphi)A_{\bar{\bar{\rho}}}. \tag{3.4}$$

Consider an antisymmetric tensor $A_{\underset{\vee}{\mu\nu},\rho}$.* Let

$$\overset{*}{A}_{\underset{\vee}{\mu\nu},\rho} = \frac{\sqrt{-\gamma}}{2}\epsilon_{\mu\nu\pi\sigma}\gamma^{\pi\lambda}\gamma^{\sigma\tau}A_{\underset{\vee}{\lambda\tau},\rho}. \tag{3.5}$$

---

* In the following, the comma (appearing in $A_{\mu\nu,\rho}$) simply separates the subscripts and has no relation to the derivative symbol.

Then

$$\underset{\vee}{A}^{**}_{\mu\nu,\rho} = \frac{\sqrt{-\gamma}}{2}\,\epsilon_{\mu\nu\pi\sigma}\,\gamma^{\pi\lambda}\,\gamma^{\sigma\tau}\,\underset{\vee}{A}^{*}_{\lambda\tau,\rho} = -\,\underset{\vee}{A}_{\mu\nu,\rho}. \tag{3.6}$$

A. The Equations $g_{\underset{+-}{\mu\nu};\rho} = 0$.

The density equations

$$\mathfrak{G}^{\underset{+-}{\mu\nu}};\rho \equiv \partial_\rho \mathfrak{G}^{\mu\nu} + \Delta^{\mu}_{\sigma\rho}\,\mathfrak{G}^{\sigma\nu} + \Delta^{\nu}_{\rho\sigma}\,\mathfrak{G}^{\mu\sigma} - \mathfrak{G}^{\mu\nu}\underline{\Delta^{\sigma}_{\rho\sigma}} = 0$$

are equivalent to either of the following systems:

$$g^{\underset{+-}{\mu\nu}};\rho = \partial_\rho\, g^{\mu\nu} + \Delta^{\mu}_{\sigma\rho}\; g^{\sigma\nu} + \Delta^{\nu}_{\rho\sigma}\; g^{\mu\sigma} = 0 \tag{3.7}$$

$$g_{\underset{+-}{\mu\nu}};\rho = \partial_\rho\, g_{\mu\nu} - \Delta^{\sigma}_{\mu\rho}\; g_{\sigma\nu} - \Delta^{\sigma}_{\rho\nu}\; g_{\mu\sigma} = 0 \tag{3.8}$$

We will solve the first group of Einstein's equations starting with this last form.

1. CALCULATION OF $\underline{\Delta^{\rho}_{\mu\rho}}$ AND $\underset{\vee}{\Delta}^{\rho}_{\mu\rho}$.

We multiply (3.8) by $g^{\mu\nu}$ and sum obtaining:

$$g^{\mu\nu}\,\partial_\rho\, g_{\mu\nu} - \Delta^{\mu}_{\mu\rho} - \Delta^{\mu}_{\rho\mu} = 0$$

or

$$\underline{\Delta^{\mu}_{\rho\mu}} = \frac{1}{2g}\,\partial_\mu g = \partial_\mu \log\sqrt{-g}. \tag{3.9}$$

We form $g^{\underset{+-}{\mu\rho}};_\rho - g^{\underset{+-}{\rho\mu}};\rho$. From (3.7), we have:

$$\frac{1}{2}\{g^{\underset{+-}{\mu\rho}};\rho - g^{\underset{+-}{\rho\mu}};\rho\} = \partial_\rho\, f^{\mu\rho} + \underline{\Delta^{\rho}_{\sigma\rho}}\, f^{\mu\sigma} - \Delta^{\rho}_{\sigma\rho}\; h^{\mu\sigma} = 0$$
$$(g^{\mu\nu} = f^{\mu\nu} + h^{\mu\nu}) \tag{3.10}$$

Let

$$\mathfrak{f}^{\mu} = \partial_\rho\,(\sqrt{-g}\; f^{\mu\rho}), \qquad f^{\mu} = \frac{1}{\sqrt{-g}}\,\partial_\rho(\sqrt{-g}\; f^{\mu\rho}). \tag{3.11}$$

We multiply (3. 10) by $h_{\mu\nu}$ and sum over μ. We obtain, using (1. 6) and (3. 9):

$$\begin{aligned}\underset{V}{\Delta}{}^{\rho}_{\nu\rho} &= h_{\mu\nu}(\partial_\rho f^{\mu\rho} + \Delta^{\rho}_{\underline{\sigma\rho}} f^{\mu\sigma}) \\ &= h_{\mu\nu}\frac{1}{\sqrt{-g}}\partial_\rho(\sqrt{-g}\, f^{\mu\rho}) = h_{\mu\nu} f^{\mu}.\end{aligned} \tag{3. 12}$$

Using (1. 20) and (3. 2), we have:

$$\underset{V}{\Delta}{}^{\rho}_{\nu\rho} = (\gamma_{\mu\nu} + \varphi_{\mu\rho}\varphi_{\nu\sigma}\gamma^{\rho\sigma}) f^{\mu} = f_\nu - \bar{f}_{\bar{\nu}} = \Delta_\nu. \tag{3. 13}$$

2. EXPRESSION FOR THE AFFINE CONNECTION AS A FUNCTION OF ITS ANTISYMMETRICAL PART.*

Consider again Eq. (3. 8). Separation into its symmetric and antisymmetric parts in μ, ν leads to:

$$(S_1)\quad \partial_\rho \gamma_{\mu\nu} - (\Delta^{\sigma}_{\underline{\mu\sigma}}\gamma_{\sigma\nu} + \Delta^{\sigma}_{\underline{\nu\rho}}\gamma_{\sigma\mu}) - (\underset{V}{\Delta}{}^{\sigma}_{\mu\rho}\varphi_{\sigma\nu} + \underset{V}{\Delta}{}^{\sigma}_{\nu\rho}\varphi_{\sigma\mu}) = 0$$

$$(A_1)\quad \partial_\rho \varphi_{\mu\nu} - (\Delta^{\sigma}_{\underline{\mu\sigma}}\varphi_{\underline{\sigma\nu}} - \Delta^{\sigma}_{\underline{\nu\rho}}\varphi_{\sigma\mu}) - (\underset{V}{\Delta}{}^{\sigma}_{\mu\rho}\gamma_{\sigma\nu} - \underset{V}{\Delta}{}^{\sigma}_{\nu\rho}\gamma_{\sigma\mu}) = 0$$

A permutation of ρ and μ and then ρ and ν leads to four additional equations analogous to the above two. Labeling these equations (S_2), (A_2), (S_3) and (A_3), we form the quantities $(S) = (S_2) + (S_3) - (S_1)$ and $(A) = (A_2) + (A_3) + (A_1)$. This leads to:

$$\partial_\mu\gamma_{\nu\rho} + \partial_\nu\gamma_{\mu\rho} - \partial_\rho\gamma_{\mu\nu} - 2\Delta^{\sigma}_{\underline{\mu\nu}}\gamma_{\sigma\rho} + 2(\underset{V}{\Delta}{}^{\sigma}_{\mu\rho}\varphi_{\sigma\nu} + \underset{V}{\Delta}{}^{\sigma}_{\nu\rho}\varphi_{\sigma\mu}) = 0. \tag{S}$$

$$\begin{aligned}\partial_\mu\varphi_{\rho\nu} + \partial_\nu\varphi_{\mu\rho} + \partial_\rho\varphi_{\mu\nu} &- 2(\Delta^{\sigma}_{\underline{\mu\rho}}\varphi_{\sigma\nu} - \Delta^{\sigma}_{\underline{\nu\rho}}\varphi_{\sigma\mu}) \\ &- 2\,\underset{V}{\Delta}{}^{\sigma}_{\mu\nu}\gamma_{\sigma\rho} = 0.\end{aligned} \tag{A}$$

---

* See (53), (54), (55) and (57).

We introduce the symbols:

$$\left.\begin{aligned}[\mu\nu,\rho] &= \frac{1}{2}(\partial_\mu \gamma_{\nu\rho} + \partial_\nu \gamma_{\mu\rho} - \partial_\rho \gamma_{\mu\nu}) \\ \left\{ {\rho \atop \mu\nu} \right\} &= \frac{1}{2}\gamma^{\rho\sigma}(\partial_\mu \gamma_{\nu\sigma} + \partial_\nu \gamma_{\mu\sigma} - \partial_\sigma \gamma_{\mu\nu})\end{aligned}\right\} \qquad (3.14)$$

and let ∇_ρ represent the covariant derivative with respect to these symbols. In particular, we have:

$$\nabla_\rho \varphi_{\mu\nu} = \partial_\rho \varphi_{\mu\nu} - \left\{ {\sigma \atop \mu\rho} \right\} \varphi_{\sigma\nu} - \left\{ {\sigma \atop \mu\nu} \right\} \varphi_{\mu\sigma}$$

and by definition $\nabla_\rho \gamma_{\mu\nu} = 0$.

If $u_{\mu\nu}^{\rho}$ represents the symmetrical part of the affine connection other than $\{\mu\nu^\rho\}$, then

$$\Delta_{\underline{\mu\nu}}^{\rho} = \left\{ {\rho \atop \mu\nu} \right\} + u_{\mu\nu}^{\rho} \qquad (3.15)$$

and, if we lower the superscript ρ of $u_{\mu\nu}^{\rho}$ and $\Delta_{\mu\nu}^{\rho}$ by letting

$$u_{\mu\nu,\rho} = \gamma_{\rho\sigma} u_{\mu\nu}^{\sigma}, \quad \Delta_{\underset{\vee}{\mu\nu},\rho} = \gamma_{\rho\sigma} \Delta_{\underset{\vee}{\mu\nu}}^{\sigma}, \qquad (3.16)$$

we note that the relations (S) and (A) can be written as

$$u_{\mu\nu,\rho} = \Delta_{\underset{\vee}{\mu\rho}}^{\sigma} \varphi_{\sigma\nu} + \Delta_{\underset{\vee}{\nu\rho}}^{\sigma} \varphi_{\sigma\mu}, \qquad (3.17)$$

$$\Delta_{\underset{\vee}{\mu\nu},\rho} = -\frac{1}{2}\varphi_{\mu\nu\rho} + \nabla_\rho \varphi_{\mu\nu} - (u_{\mu\rho}^{\sigma} \varphi_{\sigma\nu} - u_{\nu\rho}^{\sigma} \varphi_{\sigma\mu}), \qquad (3.18)$$

with

$$\varphi_{\mu\nu\rho} = \partial_\rho \varphi_{\mu\nu} + \partial_\nu \varphi_{\rho\mu} + \partial_\mu \varphi_{\nu\rho}. \qquad (3.19)$$

The equation $g_{\underset{+-}{\mu\nu};\rho} = 0$ equivalent to the systems (S) and(A) can thus be divided into the 40 equations (3. 17) which determine the symmetrical part of the affine connection and the 24 equations (3. 18) which upon substitution of (3. 17) determine the antisymmetrical part of the affine connection in terms of the field quantities. The problem thus is the determination of the antisymmetric part of the affine connection.

If we were to form $(A_2) + (A_3) - (A_1)$ and $(S_2) + (S_3) + (S_1)$, we would obtain the system

$$\partial_\mu \gamma_{\nu\rho} + \partial_\nu \gamma_{\mu\rho} + \partial_\rho \gamma_{\mu\nu} - 2(\Delta^{\sigma}_{\underline{\mu\nu}} \gamma_{\sigma\rho} + \Delta^{\sigma}_{\underline{\nu\rho}} \gamma_{\sigma\mu} + \Delta^{\sigma}_{\underline{\mu\nu}} \gamma_{\sigma\rho}) = 0 \qquad (S')$$

$$\partial_\mu \varphi_{\nu\rho} + \partial_\nu \varphi_{\mu\rho} + \partial_\rho \varphi_{\mu\nu} - 2(\underset{\vee}{\Delta^{\sigma}_{\mu\nu}} \gamma_{\sigma\rho} + \underset{\vee}{\Delta^{\sigma}_{\nu\rho}} \gamma_{\sigma\mu} + \underset{\vee}{\Delta^{\sigma}_{\mu\nu}} \gamma_{\sigma\rho}) = 0 \qquad (A')$$

which is a consequence of (S) and (A). With the notations (3. 16) and using (3. 15), we have:

$$u_{\mu\nu\rho} = u_{\mu\nu,\rho} + u_{\rho\mu,\nu} + u_{\nu\rho,\mu} = 0 \qquad (3.20)$$

$$\Delta_{\mu\nu\rho} = \underset{\vee}{\Delta_{\mu\nu,\rho}} + \underset{\vee}{\Delta_{\rho\mu,\nu}} + \underset{\vee}{\Delta_{\nu\rho,\mu}} = -\frac{1}{2} \varphi_{\mu\nu\rho}. \qquad (3.21)$$

3. THE 24 EQUATIONS IN $\underset{\vee}{\Delta^{\rho}_{\mu\nu}}$.

To solve (3. 17) and (3. 18), it is sufficient, in principle to substitute the $u^{\rho}_{\mu\nu}$ from (3. 17) into (3. 18) and solve for $\underset{\vee}{\Delta^{\rho}_{\mu\nu}}$. Starting from (3. 17), we form the expression

$u^{\sigma}_{\mu\rho} \varphi_{\sigma\nu} - u^{\sigma}_{\nu\rho} \varphi_{\sigma\mu}$ in (3. 18). We have:

$$u^{\sigma}_{\mu\rho} \varphi_{\sigma\nu} - u^{\sigma}_{\nu\rho} \varphi_{\sigma\mu}$$
$$= \gamma^{\sigma\lambda} \{ \underset{\vee}{\Delta^{\tau}_{\mu\lambda}} \varphi_{\tau\rho} \varphi_{\sigma\nu} + \underset{\vee}{\Delta^{\tau}_{\rho\lambda}} \varphi_{\tau\mu} \varphi_{\sigma\nu} - \underset{\vee}{\Delta^{\tau}_{\mu\lambda}} \varphi_{\tau\rho} \varphi_{\sigma\mu} - \Delta^{\tau}_{\rho\lambda} \varphi_{\tau\nu} \varphi_{\sigma\mu} \}. \qquad (3.22)$$

Thus, using the notations of (3. 2), we have:

$$
\begin{cases}
\underset{v}{\Delta}_{\mu\nu,\rho} = \gamma_{\rho\sigma}\, \underset{v}{\Delta}{}^{\sigma}_{\mu\nu} \\
\underset{v}{\Delta}_{\mu\nu,\bar{\rho}} = \varphi_{\rho\sigma}\, \gamma^{\sigma\tau}\, \underset{v}{\Delta}_{\mu\nu,\tau} = \varphi_{\rho\sigma}\, \underset{v}{\Delta}{}^{\sigma}_{\mu\nu}\,, \\
\underset{v}{\Delta}_{\mu\nu,\bar{\bar{\rho}}} = \varphi_{\rho\sigma}\, \gamma^{\sigma\tau} \underset{v}{\Delta}_{\mu\nu,\bar{\tau}} = \varphi_{\rho\sigma}\, \gamma^{\sigma\tau}\, \varphi_{\tau\lambda}\, \underset{v}{\Delta}{}^{\lambda}_{\mu\nu}\,.
\end{cases}
\tag{3.23}
$$

Let

$$
\begin{cases}
\Delta_{\mu\nu\rho} = \underset{v}{\Delta}_{\mu\nu,\rho} + \underset{v}{\Delta}_{\rho\mu,\nu} + \underset{v}{\Delta}_{\nu\rho,\mu} \\
\bar{\Delta}_{\mu\nu\rho} = \underset{v}{\Delta}_{\mu\nu,\bar{\rho}} + \underset{v}{\Delta}_{\rho\mu,\bar{\nu}} + \underset{v}{\Delta}_{\nu\rho,\bar{\mu}} \\
\bar{\bar{\Delta}}_{\mu\nu\rho} = \underset{v}{\Delta}_{\mu\nu,\bar{\bar{\rho}}} + \underset{v}{\Delta}_{\rho\mu,\bar{\bar{\nu}}} + \underset{v}{\Delta}_{\nu\rho,\bar{\bar{\mu}}}\,.
\end{cases}
\tag{3.24}
$$

We then obtain from (3. 22) (see (57), App. III):

$$
\begin{aligned}
u^{\sigma}_{\mu\rho}\;\varphi_{\sigma\nu} - u^{\sigma}_{\nu\rho}\;\varphi_{\sigma\mu} = &- 2\gamma^{\sigma\lambda}[\underset{v}{\Delta}_{\rho\lambda,\bar{\mu}}\,\varphi_{\sigma\nu} - \underset{v}{\Delta}_{\rho\lambda,\bar{\nu}}\,\varphi_{\sigma\mu}] \\
&- \gamma^{\sigma\lambda}[\bar{\Delta}_{\mu\lambda\rho}\,\varphi_{\sigma\nu} - \bar{\Delta}_{\nu\lambda\rho}\,\varphi_{\sigma\mu}] \\
&- \bar{\bar{\Delta}}_{\mu\nu\rho} + \underset{v}{\Delta}_{\mu\nu,\bar{\bar{\rho}}}\,.
\end{aligned}
\tag{3.25}
$$

Substitution of (3. 25) in (3. 18) leads to:

$$
\begin{aligned}
&\underset{v}{\Delta}_{\mu\nu,\rho} - \gamma^{\sigma\lambda}(\bar{\Delta}_{\mu\lambda\rho}\,\varphi_{\sigma\nu} - \bar{\Delta}_{\nu\lambda\rho}\,\varphi_{\sigma\mu}) \\
&\quad - 2\gamma^{\sigma\lambda}(\underset{v}{\Delta}_{\rho\lambda,\bar{\mu}}\,\varphi_{\sigma\nu} - \underset{v}{\Delta}_{\rho\lambda,\bar{\nu}}\,\varphi_{\sigma\mu}) + \underset{v}{\Delta}_{\mu\nu,\bar{\bar{\rho}}} - \bar{\bar{\Delta}}_{\mu\nu\rho} \\
&\quad = -\frac{1}{2}\,\varphi_{\mu\nu\rho} + \Delta_{\rho}\,\varphi_{\mu\nu}\,.
\end{aligned}
\tag{3.26}
$$

It is then sufficient to solve these 24 equations to entirely determine the affine connection.

4. DERIVATION OF THE 24 EQUATIONS IN $\underset{\vee}{\Delta}_{\mu\nu,\rho}$, $\overset{*}{\Delta}_{\mu\nu,\rho}$ AND $\Delta_{\mu\nu,\bar{\bar{\rho}}}$.

We shall transform (3. 26) in such a way that the 24 antisymmetric coefficients $\underset{\vee}{\Delta}^{\rho}_{\mu\nu}$ will be represented by the following terms:

$$\underset{\vee}{\Delta}_{\mu\nu,\rho} = \gamma_{\rho\sigma}\,\underset{\vee}{\Delta}^{\sigma}_{\mu\nu} \tag{3. 27}$$

$$\Delta_{\mu\nu,\bar{\bar{\rho}}} = \varphi_{\rho\sigma}\,\gamma^{\sigma\tau}\,\underset{\vee}{\Delta}_{\mu\nu,\bar{\tau}} = \varphi_{\rho\sigma}\,\gamma^{\sigma\tau}\,\varphi_{\tau\lambda}\,\underset{\vee}{\Delta}^{\lambda}_{\mu\nu}\,, \tag{3. 28}$$

$$\overset{*}{\underset{\vee}{\Delta}}_{\mu\nu,\rho} = \frac{\sqrt{-\gamma}}{2}\,\epsilon_{\mu\nu\epsilon\sigma}\,\gamma^{\epsilon\lambda}\,\gamma^{\sigma\tau}\,\underset{\vee}{\Delta}_{\lambda\tau,\rho}. \tag{3. 29}$$

To this end, we shall introduce the intermediate notations:

$$A_{\rho} = \frac{1}{2}\,\gamma^{\mu\lambda}\,\gamma^{\nu\tau}\,\varphi_{\mu\nu}\,\underset{\vee}{\Delta}_{\lambda\tau,\rho} \tag{3. 30}$$

$$B_{\rho} = \frac{1}{2}\,\varphi^{\mu\nu}\,\underset{\vee}{\Delta}_{\mu\nu,\rho}. \tag{3. 31}$$

We must compute the three expressions

$$\bar{\Delta}_{\mu\nu\rho},\ \gamma^{\sigma\lambda}(\underset{\vee}{\Delta}_{\rho\lambda,\bar{\mu}}\,\varphi_{\sigma\nu} - \underset{\vee}{\Delta}_{\rho\lambda,\bar{\nu}}\,\varphi_{\sigma\mu}),\ \text{and}\ \bar{\bar{\Delta}}_{\mu\nu\rho},$$

which appear in (3. 26) as a function of the quantities given in (3. 27) through (3. 31) (with the same μ, ν, ρ indices in the same equation). After some calculations (see (57), Notes IV, V, VI), one obtains:

$$\bar{\Delta}_{\mu\lambda\rho} = -(\varphi_{\mu\lambda}\,\Delta_{\rho} + \varphi_{\rho\mu}\,\Delta_{\lambda} + \varphi_{\lambda\rho}\,\Delta_{\mu}) + \sqrt{\varphi}\,\epsilon_{\mu\lambda\rho\sigma}\,\gamma^{\sigma\tau}B_{\tau} \tag{3. 32}$$

$$
\begin{aligned}
&\gamma^{\sigma\lambda}(\varphi_{\sigma\nu}\,\Delta_{\underset{\vee}{\rho\lambda},\bar{\mu}} - \varphi_{\sigma\mu}\,\Delta_{\underset{\vee}{\rho\lambda},\bar{\nu}}) \\
&= -\frac{1}{2}\gamma^{\sigma\lambda}\,\gamma^{\tau\pi}(\sqrt{\varphi}\,\epsilon_{\mu\nu\sigma\tau} - \varphi_{\sigma\tau}\,\varphi_{\mu\nu})\Delta_{\rho\lambda\pi} \\
&\quad + \frac{\sqrt{\varphi}}{\sqrt{-\gamma}}\,\overset{*}{\Delta}_{\underset{\vee}{\mu\nu},\rho} - \varphi_{\mu\nu}\,A_\rho,
\end{aligned}
\tag{3.33}
$$

$$
\begin{aligned}
\bar{\bar{\Delta}}_{\mu\nu\rho} = {} & \left[1 - \frac{g}{\gamma} + \frac{\varphi}{\gamma}\right]\Delta_{\mu\nu\rho} + \sqrt{\varphi}\,\epsilon_{\mu\nu\rho\sigma}\,\varphi^{\lambda\sigma}A_\lambda \\
& + \sqrt{\varphi}\,\epsilon_{\mu\nu\rho\sigma}\,\gamma^{\lambda\alpha}\,\gamma^{\sigma\beta}\varphi_{\alpha\beta}B_\lambda \\
& + \frac{\sqrt{\varphi}}{2}\,\epsilon_{\mu\nu\rho\lambda}\gamma^{\alpha\sigma}\,\gamma^{\lambda\tau}\,\varphi_{\sigma\tau}\,\varphi^{\beta\delta}\,\Delta_{\alpha\beta\delta}.
\end{aligned}
\tag{3.34}
$$

Substitution of (3. 32), (3. 33) and (3. 34) into (3. 26) leads to:

$$
\Delta_{\underset{\vee}{\mu\nu},\rho} + \Delta_{\underset{\vee}{\mu\nu},\bar{\bar{\rho}}} - \frac{2\sqrt{\varphi}}{\sqrt{-\gamma}}\,\overset{*}{\Delta}_{\underset{\vee}{\mu\nu},\rho}
\tag{3.35}
$$

$$
\begin{aligned}
&= -\frac{1}{2}\varphi_{\mu\nu\rho} + \nabla_\rho\,\varphi_{\mu\nu} - \gamma^{\sigma\lambda}(\varphi_{\rho\mu}\,\varphi_{\sigma\nu} - \varphi_{\rho\nu}\varphi_{\sigma\mu} - \varphi_{\rho\sigma}\,\varphi_{\mu\nu})\Delta_\lambda \\
&\quad -\gamma^{\sigma\lambda}\,\varphi_{\lambda\rho}(\varphi_{\sigma\nu}\Delta_\mu + \varphi_{\mu\sigma}\,\Delta_\nu + \varphi_{\nu\mu}\,\Delta_\sigma) \\
&\quad -\gamma^{\sigma\lambda}\,\gamma^{\tau\pi}(\sqrt{\varphi}\,\epsilon_{\mu\nu\sigma\tau} - \varphi_{\sigma\tau}\,\varphi_{\mu\nu})\,\Delta_{\rho\lambda\pi} \\
&\quad + \left(1 - \frac{g}{\gamma} + \frac{\varphi}{\gamma}\right)\Delta_{\mu\nu\rho} + \frac{\sqrt{\varphi}}{2}\,\epsilon_{\mu\nu\rho\lambda}\gamma^{\alpha\sigma}\,\gamma^{\lambda\tau}\,\varphi_{\sigma\tau}\,\varphi^{\beta\delta}\,\Delta_{\alpha\beta\delta} \\
&\quad - \sqrt{\varphi}\,\epsilon_{\mu\nu\lambda\pi}\gamma^{\lambda\sigma}\,\gamma^{\pi\tau}\,\varphi_{\sigma\rho}\,B_\tau + \sqrt{\varphi}\,\epsilon_{\mu\nu\rho\sigma}\,\varphi^{\lambda\sigma}\,A_\lambda - 2\varphi_{\mu\nu}\,A_\rho.
\end{aligned}
$$

To obtain the desired equations, we must compute A_ρ and B_ρ.

a. Calculation of B_ρ:* We multiply (3. 18) by $\frac{1}{2}\varphi_{\mu\nu}$ and contract obtaining:

$$B_\rho = -\frac{1}{4}\varphi^{\mu\nu}\varphi_{\mu\nu\rho} + \frac{1}{2}\varphi^{\mu\nu}\nabla_\rho\varphi_{\mu\nu} - u_\rho \tag{3. 36}$$

But

$$\varphi^{\mu\nu}\nabla_\rho\varphi_{\mu\nu} = \frac{1}{\varphi}\partial_\rho\varphi - \frac{1}{\gamma}\partial_\rho\gamma = \partial_\rho \operatorname{Log}\frac{\varphi}{\gamma}. \tag{3. 37}$$

and from (3. 9), we had

$$\Delta^\rho_{\underline{\mu\rho}} = \left\{{\rho \atop \mu\rho}\right\} + u^\rho_{\mu\rho} = \frac{1}{2}\partial_\mu \operatorname{Log} g \tag{3. 38}$$

or

$$u_\mu = u^\rho_{\mu\rho} = \frac{1}{2}\partial_\mu \operatorname{Log}\frac{g}{\gamma}. \tag{3. 39}$$

Thus, upon substitution in (3. 36), we have:

$$B_\rho = -\frac{1}{4}\varphi^{\mu\nu}\varphi_{\mu\nu\rho} + \frac{1}{2}\partial_\rho \operatorname{Log}\frac{\varphi}{g}. \tag{3. 40}$$

b. Computation of A_ρ: We multiply (3. 26) by $\frac{1}{2}\varphi^{\mu\lambda}$ and contract. This leads to (see (57), Note VII):

$$\begin{aligned} B_\rho + A_\rho - \frac{1}{2}\gamma^{\alpha\mu}\gamma^{\beta\nu}\varphi_{\alpha\beta}\Delta_{\mu\nu\rho} \\ = -\frac{1}{4}\varphi^{\mu\nu}\varphi_{\mu\nu\rho} + \frac{1}{2}\varphi^{\mu\nu}\nabla_\rho\varphi_{\mu\nu}. \end{aligned} \tag{3. 41}$$

or using (3. 21) and (3. 37):

$$A_\rho = -\frac{1}{4}\gamma^{\alpha\mu}\gamma^{\beta\nu}\varphi_{\alpha\beta}\varphi_{\mu\nu\rho} + \frac{1}{2}\partial_\rho \operatorname{Log}\frac{g}{\gamma}.$$

---

* It might seem preferable to compute

$$B'_\rho = \sqrt{\varphi}\, B_\rho = \tfrac{1}{4}\epsilon^{\mu\nu\lambda\sigma}\varphi_{\lambda\sigma}\Delta_{\mu\nu,\rho}$$

since B'_ρ has a meaning when $\varphi = 0$ whereas B_ρ does not. The results obtained are equivalent since it is B'_ρ and not B_ρ which occurs in (3.32) and (3.34).

One can now substitute A_ρ and B_ρ in (3. 35). After some computations (see (57), note VIII), one can write

$$\Delta_{\underset{\vee}{\mu\nu},\rho} - \frac{2\sqrt{\varphi}}{\sqrt{-\gamma}} \overset{*}{\Delta}_{\underset{\vee}{\mu\nu},\rho} + \Delta_{\underset{\vee}{\mu\nu},\bar{\bar{\rho}}} = R_{\underset{\vee}{\mu\nu},\rho} \tag{3. 43}$$

where *

$$R_{\underset{\vee}{\mu\nu},\rho} = -\frac{1}{2}\varphi_{\mu\nu\rho} + \nabla_\rho \varphi_{\mu\nu} + \frac{\sqrt{\varphi}}{2\sqrt{-\gamma}}\overset{*}{\varphi}_{[\mu\nu]\rho} + \frac{\sqrt{\varphi}}{4\sqrt{-\gamma}}\overset{*}{\varphi}_{\mu\nu}\varphi^{\sigma\tau}\varphi_{\sigma\tau\rho}$$

$$-\varphi_{\mu\nu}\partial_\rho \operatorname{Log}\frac{g}{\gamma} + \frac{\sqrt{\varphi}}{2}\epsilon_{\mu\nu\rho\sigma}\varphi^{\lambda\sigma}\partial_\lambda \operatorname{Log}\frac{g}{\gamma} \tag{3. 44}$$

$$-\frac{\varphi}{2\sqrt{-\gamma}}\overset{*}{\epsilon}_{[\mu\nu]\rho\lambda}\varphi^{\sigma\lambda}\partial_\sigma \operatorname{Log}\frac{g}{\varphi} + \frac{\sqrt{\varphi}}{2\sqrt{-\gamma}}\overset{*}{\varphi}_{\mu\nu}\partial_\rho \operatorname{Log}\frac{g}{\varphi}$$

$$+\gamma^{\sigma\lambda}[\sqrt{\varphi}\,\epsilon_{\mu\nu\rho\sigma}\Delta_\lambda + \varphi_{\lambda\rho}(\varphi_{\mu\nu}\Delta_\sigma + \varphi_{\sigma\mu}\Delta_\nu + \varphi_{\nu\sigma}\Delta_\mu)].$$

All the terms occurring in $R_{\underset{\vee}{\mu\nu},\rho}$ are determined in terms of $\gamma_{\mu\nu}$ and $\varphi_{\mu\nu}$. We recall that ∇_ρ, the covariant derivative in terms of $\gamma_{\mu\nu}$, is a function of γ. The Δ_ρ are known from (3. 13):

$$\Delta_\rho = f_\rho - f_{\bar{\bar{\rho}}}.$$

Further from (3. 5):

$$\overset{*}{\varphi}_{\mu\nu} = \frac{1}{2}\sqrt{-\gamma}\,\epsilon_{\mu\nu\alpha\beta}\gamma^{\alpha\lambda}\gamma^{\beta\sigma}\varphi_{\lambda\sigma}.$$

$$\overset{*}{\varphi}_{[\mu\nu]\rho} = \frac{1}{2}\sqrt{-\gamma}\,\epsilon_{\mu\nu\alpha\beta}\gamma^{\alpha\lambda}\gamma^{\beta\sigma}\varphi_{\lambda\sigma\rho}.$$

$$\overset{*}{\epsilon}_{[\mu\nu]\rho\lambda} = \frac{1}{2}\sqrt{-\gamma}\,\epsilon_{\mu\nu\alpha\beta}\gamma^{\alpha\tau}\gamma^{\beta\sigma}\epsilon_{\tau\sigma\rho\lambda}.$$

---

* We must bear in mind that the letter R used here (always designated by three indices) cannot be confused with $R_{\mu\nu}$, Ricci's tensor (always two indices). We also recall that the comma represents a lowering of an index and has no relation with the ordinary derivative.

5. SOLUTION OF THE EQUATION $\Delta_{\underset{\vee}{\mu\nu},\rho} - 2\sqrt{\varphi}/\sqrt{-\gamma}\ \overset{*}{\Delta}_{\underset{\vee}{\mu\nu},\rho} + \Delta_{\underset{\vee}{\mu\nu},\bar{\bar{\rho}}} = R_{\underset{\vee}{\mu\nu},\rho}$.

Let $(\mathcal{E}')$ andresent the relation $\mathcal{E}_{\underset{\vee}{\mu\nu},\rho} = 0$ corresponding to (3.43) and $(\mathcal{E}^*)$ and $(\bar{\bar{\mathcal{E}}})$ the equations corresponding to $\mathcal{E}^*_{\underset{\vee}{\mu\nu},\rho}$ and $\mathcal{E}_{\underset{\vee}{\mu\nu},\bar{\bar{\rho}}}$. These equations can be deduced from $\mathcal{E}_{\underset{\vee}{\mu\nu},\rho}$ in the usual way:

$$\overset{*}{\mathcal{E}}_{\underset{\vee}{\mu\nu},\rho} = \frac{1}{2}\sqrt{-\gamma}\ \epsilon_{\mu\nu\alpha\beta}\,\gamma^{\alpha\lambda}\,\gamma^{\beta\sigma}\,\mathcal{E}_{\underset{\vee}{\lambda\sigma},\rho}. \tag{3.45}$$

$$\mathcal{E}_{\underset{\vee}{\mu\nu},\bar{\bar{\rho}}} = \varphi_{\rho\sigma}\,\gamma^{\sigma\tau}\,\varphi_{\tau\lambda}\,\gamma^{\lambda\pi}\,\mathcal{E}_{\underset{\vee}{\mu\nu},\pi}. \tag{3.46}$$

Let us now form $(\mathcal{E}^*)$. Recalling that from (3.6) $\overset{**}{\Delta}_{\underset{\vee}{\mu\nu},\rho} = -\Delta_{\underset{\vee}{\mu\nu},\rho}$, we have:

$$\overset{*}{\Delta}_{\mu\nu,\rho} + \frac{2\sqrt{\varphi}}{\sqrt{-\gamma}}\Delta_{\underset{\vee}{\mu\nu},\rho} + \overset{*}{\Delta}_{\underset{\vee}{\mu\nu},\rho} = \overset{*}{R}_{\mu\nu,\rho}. \tag{$\mathcal{E}^*$}$$

and

$$\Delta_{\underset{\vee}{\mu\nu},\bar{\bar{\rho}}} - \frac{2\sqrt{\varphi}}{\sqrt{-\gamma}}\ \overset{*}{\Delta}_{\underset{\vee}{\mu\nu},\bar{\bar{\rho}}} + \Delta_{\mu\nu,\bar{\bar{\bar{\rho}}}} = R_{\mu\nu,\bar{\bar{\rho}}}. \tag{$\bar{\bar{\mathcal{E}}}$}$$

Thus the quantity $2\sqrt{\varphi}/\sqrt{-\gamma}\ (\mathcal{E}^*) + (\bar{\bar{\mathcal{E}}})$ yields:

$$\begin{aligned}\frac{2\sqrt{\varphi}}{\sqrt{-\gamma}}\ \overset{*}{\Delta}_{\underset{\vee}{\mu\nu},\rho} - \frac{4\varphi}{\gamma}\ \Delta_{\underset{\vee}{\mu\nu},\rho} + \Delta_{\underset{\vee}{\mu\nu},\bar{\bar{\rho}}} + \Delta_{\underset{\vee}{\mu\nu},\bar{\bar{\bar{\rho}}}} \\ = \frac{2\sqrt{\varphi}}{\sqrt{-\gamma}}\ \overset{*}{R}_{\underset{\vee}{\mu\nu},\rho} + R_{\underset{\vee}{\mu\nu},\bar{\bar{\rho}}}.\end{aligned} \tag{3.47}$$

If we apply (3.4) to $\Delta_{\underset{\vee}{\mu\nu},\rho}$, we have:

$$\Delta_{\underset{\vee}{\mu\nu},\bar{\bar{\bar{\rho}}}} = -\frac{\varphi}{\gamma}\Delta_{\underset{\vee}{\mu\nu},\rho} + \left(1 - \frac{g}{\gamma} + \frac{\varphi}{\gamma}\right)\Delta_{\underset{\vee}{\mu\nu},\bar{\bar{\rho}}}. \tag{3.48}$$

Comparison with (3.47) leads to:

$$\frac{2\sqrt{\varphi}}{\sqrt{-\gamma}}\overset{*}{\Delta}_{\underset{\vee}{\mu\nu},\rho} - \frac{5\varphi}{\gamma}\Delta_{\underset{\vee}{\mu\nu},\rho} + \left(2 - \frac{g}{\gamma} + \frac{\varphi}{\gamma}\right)\Delta_{\underset{\vee}{\mu\nu},\bar{\bar{\rho}}}$$
$$= \frac{2\sqrt{\varphi}}{\sqrt{-\gamma}}\overset{*}{R}_{\underset{\vee}{\mu\nu},\rho} + R_{\underset{\vee}{\mu\nu},\bar{\bar{\rho}}}. \tag{3.49}$$

Subtraction of the quantity $(2 - g/\gamma + \varphi/\gamma)\,(\mathcal{E})$ leads to:

$$\left(2 - \frac{g}{\gamma} + \frac{6\varphi}{\gamma}\right)\Delta_{\underset{\vee}{\mu\nu},\rho} - \frac{2\sqrt{\varphi}}{\sqrt{-\gamma}}\left(3 - \frac{g}{\gamma} + \frac{\varphi}{\gamma}\right)\overset{*}{\Delta}_{\underset{\vee}{\mu\nu},\rho}$$
$$= \left(2 - \frac{g}{\gamma} + \frac{\varphi}{\gamma}\right)R_{\underset{\vee}{\mu\nu},\rho} - \frac{2\sqrt{\varphi}}{\sqrt{-\gamma}}\overset{*}{R}_{\underset{\vee}{\mu\nu},\rho} - R_{\underset{\vee}{\mu\nu},\bar{\bar{\rho}}}. \tag{3.50}$$

We can form (3.50)* and then compute the quantity

$$\left[\left(2 - \frac{g}{\gamma} + \frac{6\varphi}{\gamma}\right)(3.50) + \frac{2\sqrt{\varphi}}{\sqrt{-\gamma}}\left(3 - \frac{g}{\gamma} + \frac{\varphi}{\gamma}\right)(3.50)^*\right].$$

This yields:

$$\left[\left(2 - \frac{g}{\gamma} + \frac{6\varphi}{\gamma}\right)^2 - \frac{4\varphi}{\gamma}\left(3 - \frac{g}{\gamma} + \frac{\varphi}{\gamma}\right)^2\right]\Delta_{\underset{\vee}{\mu\nu},\rho} \tag{3.51}$$
$$= \left[\left(2 - \frac{g}{\gamma} + \frac{6\varphi}{\gamma}\right)\left(2 - \frac{g}{\gamma} + \frac{\varphi}{\gamma}\right) - \frac{4\varphi}{\gamma}\left(3 - \frac{g}{\gamma} + \frac{\varphi}{\gamma}\right)\right]R_{\underset{\vee}{\mu\nu},\rho}$$
$$- \frac{2\sqrt{\varphi}}{\sqrt{-\gamma}}\left[\left(2 - \frac{g}{\gamma} + \frac{6\varphi}{\gamma}\right) - \left(2 - \frac{g}{\gamma} + \frac{\varphi}{\gamma}\right)\right]\overset{*}{R}_{\underset{\vee}{\mu\nu},\rho}$$
$$- \left(2 - \frac{g}{\gamma} + \frac{6\varphi}{\gamma}\right)R_{\underset{\vee}{\mu\nu},\bar{\bar{\rho}}} - 2\frac{\sqrt{\varphi}}{\sqrt{-\gamma}}\left(3 - \frac{g}{\gamma} + \frac{\varphi}{\gamma}\right)\overset{*}{R}_{\underset{\vee}{\mu\nu},\bar{\bar{\rho}}}.$$

This relation determines completely the antisymmetric part of the affine connection $\Delta_{\underset{\vee}{\mu\nu},\rho}$ in terms of the field quantities $g_{\mu\nu}, \gamma_{\mu\nu}, \varphi_{\mu\nu}$. The $R_{\underset{\vee}{\mu\nu},\rho}$ are uniquely determined as a function of these quantities from (3.43).

By setting

$$a = 2 - \frac{g}{\gamma} + \frac{6\varphi}{\gamma} \qquad b = 2\frac{\sqrt{\varphi}}{\sqrt{-\gamma}}\left(3 - \frac{g}{\gamma} + \frac{\varphi}{\gamma}\right) \tag{3.52}$$

and

$$S_{\mu\nu,\rho} = \left(2 - \frac{g}{\gamma} + \frac{\varphi}{\gamma}\right) R_{\mu\nu,\rho} - \frac{2\sqrt{\varphi}}{\sqrt{-\gamma}} \overset{*}{R}_{\underset{V}{\mu\nu},\rho} - R_{\underset{V}{\mu\nu},\bar{\bar{\rho}}}, \quad (3.53)$$

(3.51) can then be written in the form:

$$(a^2 + b^2)\Delta_{\underset{V}{\mu\nu},\rho} = a\, S_{\underset{V}{\mu\nu},\rho} + b\, \overset{*}{S}_{\underset{V}{\mu\nu},\rho}.$$

In summary, from the fundamental equation $g_{\underset{+-}{\mu\nu};\rho} = 0$

1° The 64 coefficients defining the affine connection can be expressed in terms of the antisymmetric part of the connection by the relation

$$\Delta^{\rho}_{\mu\nu} = \left\{ {\rho \atop \mu\nu} \right\} + u^{\rho}_{\mu\nu} + \Delta^{\rho}_{\mu\nu} \quad (3.55)$$

$$= \left\{ {\rho \atop \mu\nu} \right\} + \gamma^{\rho\lambda}\{\Delta^{\sigma}_{\underset{V}{\mu\nu}}\, \varphi_{\sigma\nu} + \Delta^{\sigma}_{\underset{V}{\nu\lambda}}\, \varphi_{\sigma\mu} + \Delta_{\underset{V}{\mu\nu},\lambda}\}.$$

2° The antisymmetric part of the connection can be expressed in terms of the $R_{\underset{V}{\mu\nu},\rho}$ by (3.51) or (3.54). $R_{\underset{V}{\mu\nu},\rho}$ is determined completely by (3.44) in terms of the fields $g_{\mu\nu}$, $\gamma_{\mu\nu}$ and $\varphi_{\mu\nu}$:

$$R_{\mu\nu,\rho} = -\frac{1}{2}\varphi_{\mu\nu\rho} + \nabla_{\rho}\varphi_{\mu\nu} + \frac{\sqrt{\varphi}}{2\sqrt{-\gamma}} \overset{*}{\varphi}_{[\mu\nu]\rho} + \frac{\sqrt{\varphi}}{4\sqrt{-\gamma}} \overset{*}{\varphi}_{\mu\nu}\varphi^{\sigma\tau}\varphi_{\sigma\tau\rho}$$

$$- \varphi_{\mu\nu}\,\partial_{\rho}\mathrm{Log}\frac{g}{\gamma} + \frac{\sqrt{\varphi}}{2}\epsilon_{\mu\nu\rho\sigma}\varphi^{\lambda\sigma}\partial_{\lambda}\mathrm{Log}\frac{g}{\gamma} \quad (3.56)$$

$$- \frac{\varphi}{2\sqrt{-\gamma}} \overset{*}{\epsilon}_{[\mu\nu]\rho\lambda}\varphi^{\sigma\lambda}\partial_{\sigma}\mathrm{Log}\frac{g}{\varphi} + \frac{\sqrt{\varphi}}{2\sqrt{-\gamma}} \overset{*}{\varphi}_{\mu\nu}\,\partial_{\rho}\mathrm{Log}\frac{g}{\varphi}$$

$$+ \gamma^{\sigma\lambda}\{\sqrt{\varphi}\,\epsilon_{\mu\nu\rho\sigma}(f_{\lambda} - f_{\bar{\bar{\lambda}}}) + \varphi_{\lambda\rho}[\varphi_{\mu\nu}(f_{\sigma} - f_{\bar{\bar{\sigma}}}) + \varphi_{\sigma\mu}(f_{\nu} - f_{\bar{\bar{\nu}}})$$
$$+ \varphi_{\nu\sigma}(f_{\mu} - f_{\bar{\bar{\mu}}})]\}$$

If one wants to determine the connection $L^{\rho}_{\underset{V}{\mu\nu}}$ such that $L_{\rho} = L^{\mu}_{\underset{V}{\rho\mu}} = 0$, all the above relations hold but the last line of $R_{\mu\nu,\rho}$ in (3.56) must be omitted ($\Delta_{\rho} = f_{\rho} - f_{\bar{\bar{\rho}}}$).

6. CONDITIONS FOR THE EXISTENCE OF SOLUTIONS.

These are evident. We always assume $\gamma \neq 0$. From (3.54), $\Delta_{\underset{\vee}{\mu\nu},\rho}$ can be expressed uniquely in terms of $R_{\underset{\vee}{\mu\nu},\rho}$ if

$$a^2 + b^2 \neq 0.$$

On the other hand the $R_{\mu\nu,\rho}$ are completely determined if

$$g \neq 0$$

We must thus have

$$g(a^2 + b^2) \neq 0 \tag{3.57}$$

or

$$g\left[\left(2 - \frac{g}{\gamma} + \frac{6\varphi}{\gamma}\right)^2 - \frac{4\varphi}{\gamma}\left(3 - \frac{g}{\gamma} + \frac{\varphi}{\gamma}\right)^2\right] \neq 0 \tag{3.58}$$

7. THE CASE $\varphi = 0$ (56).

In the solution of $g_{\underset{+-}{\mu\nu};\rho} = 0$, we never required the condition $\varphi \neq 0$. The terms in $\varphi^{\mu\nu}$ which have no meaning when $\varphi = 0$ seem to enter into our answers through B_ρ and appear in (3.43) or (3.56). But in fact, we are always dealing with quantities of the form $\sqrt{\varphi}\ \varphi^{\mu\nu}$ or $\varphi\ \varphi^{\mu\nu}$ which represent expressions of the form $\frac{1}{2}\ \epsilon^{\mu\nu\rho\sigma}\ \varphi_{\rho\sigma}$ or $\sqrt{\varphi}/2\ \epsilon^{\mu\nu\rho\sigma}\ \varphi_{\rho\sigma}$. These expressions are finite (eventually zero) when $\varphi = 0$. Similarly terms of the form $\sqrt{\varphi}\ \partial_\rho \operatorname{Log}\varphi$ represent expressions in $2\partial_\rho\sqrt{\varphi}$ and vanish.

From (3.52) and (3.54), the general solution of $g_{\underset{+-}{\mu\nu};\rho} = 0$ in the case $\varphi = 0$ reduces to

$$a\ \Delta_{\underset{\vee}{\mu\nu},\rho} = S_{\underset{\vee}{\mu\nu},\rho} = a\ R_{\underset{\vee}{\mu\nu},\rho} - R_{\underset{\vee}{\mu\nu},\overset{=}{\rho}} \tag{3.59}$$

with

$$R_{\underset{\vee}{\mu\nu},\rho} = \frac{1}{2}\varphi_{\mu\nu\rho} + \nabla_\rho\varphi_{\mu\nu} + \frac{1}{8\sqrt{-\gamma}}\overset{*}{\varphi}_{\mu\nu}\epsilon^{\sigma\tau\lambda\pi}\varphi_{\lambda\pi}\varphi_{\sigma\tau\rho}$$

$$- \varphi_{\mu\nu}\partial_\rho \log\frac{g}{\gamma} + \frac{1}{2}(\varphi_{\mu\nu}\partial_\rho + \varphi_{\rho\mu}\partial_\nu + \varphi_{\nu\rho}\partial_\mu)\mathrm{Log}\frac{g}{\gamma}$$

$$+ \gamma^{\sigma\lambda}\varphi_{\lambda\rho}[\varphi_{\mu\nu}(f_\sigma - f_{\bar{\bar{\sigma}}}) + \varphi_{\sigma\mu}(f_\nu - f_{\bar{\bar{\nu}}}) + \varphi_{\nu\sigma}(f_\mu - f_{\bar{\bar{\mu}}})] \tag{3.60}$$

the last line of $R_{\underset{\vee}{\mu\nu},\rho}$ disappears if f_ρ, and hence Δ_ρ, is zero.

If $g \neq 0$, we must assume that

$$a = 2 - \frac{g}{\gamma} \neq 0 \tag{3.61}$$

or

$$1 - \frac{1}{2}\gamma^{\mu\rho}\gamma^{\nu\sigma}\varphi_{\mu\nu}\varphi_{\rho\sigma} \neq 0 \tag{3. 62}$$

In the case of $\varphi = 0$, the affine connection can thus be obtained directly from the general solution of the equations.

8. INTEGRATIBILITY CONDITIONS (24), (28).

The equations (3. 7)

$$g^{\underset{+-}{\mu\nu}}{}_{;\rho} \equiv \partial_\rho g^{\mu\nu} + \Delta^\mu_{\sigma\rho} g^{\sigma\nu} + \Delta^\nu_{\rho\sigma} g^{\mu\sigma} = 0 \tag{3. 7}$$

imply certain conditions which can be obtained by forming

$$(\partial_\rho\partial_\sigma - \partial_\sigma\partial_\rho) g^{\mu\nu} = 0 \tag{3.63}$$

Schrödinger (23) and Bose (28) have obtained these conditions explicitly by substituting in (3. 63) the expressions for $\partial_\sigma g^{\mu\nu}$ and $\partial_\rho g^{\mu\nu}$ obtained from (3. 7). One has then

$$R^\mu_{\lambda\sigma\rho} g^{\lambda\mu} + \tilde{R}^\nu_{\lambda\sigma\rho} g^{\mu\lambda} = 0 \tag{3.64}$$

where

$$R^\mu_{\lambda\sigma\rho} = \partial_\rho\Delta^\mu_{\lambda\sigma} - \partial_\sigma\Delta^\mu_{\lambda\rho} + \Delta^\tau_{\lambda\sigma}\Delta^\mu_{\tau\rho} - \Delta^\tau_{\lambda\rho}\Delta^\mu_{\tau\sigma}. \tag{3.65}$$

is the curvature tensor formed with the Δ's and $\tilde{R}$ is the same tensor formed with $\tilde{\Delta}$.

The identities (3. 64) (when (3.7) are satisfied) can also be written,

$$R_{\mu\sigma\lambda}^{\rho}\, g_{\rho\nu} + \tilde{R}_{\nu\sigma\lambda}^{\rho}\, g_{\mu\rho} = 0 \qquad (3.66)$$

Multiplying (3. 64) by $g_{\mu\nu}$ and summing, we have

$$R_{\mu\sigma\rho}^{\mu} + \tilde{R}_{\mu\sigma\rho}^{\mu} = 0 \qquad (3.67)$$

that is

$$\partial_\rho\, \Delta_{\underline{\mu\sigma}}^{\mu} - \partial_\sigma\, \Delta_{\underline{\mu\rho}}^{\mu} = 0. \qquad (3.68)$$

It can be easily verified that this is immediately satisfied by (3. 9)

$$\Delta_{\underline{\mu\rho}}^{\mu} = \partial_\rho \operatorname{Log}\sqrt{-g}$$

Contracting ρ and μ in (3. 64) and then σ and ν, we have:

$$R_{\lambda\nu\mu}^{\mu}\, g^{\lambda\nu} + \tilde{R}_{\lambda\mu\nu}^{\mu}\, g^{\nu\lambda} = 0 \qquad (3.69)$$

or

$$(R_{\mu\nu} - \tilde{R}_{\nu\mu})\, g^{\mu\nu} = 0 \qquad (3.70)$$

These conditions can also be written as

$$g^{\mu\nu}\{(\partial_\mu\, \Delta_\nu + \partial_\nu\, \Delta_\mu) - 2\, \Delta_{\mu\nu}^{\sigma}\, \Delta_\sigma\} = 0 \qquad (3.71)$$

These are evidently satisfied for all connections $L_{\mu\nu}^{\rho}$ such that $L_\rho = 0$. (i. e., starting from $\mathfrak{H} = \mathcal{G}_{\rho}^{\mu\nu}\, R_{\mu\nu}$). One can also verify that they are satisfied when $\Delta_{\mu\nu}^{\rho}$ is any connection such that $\mathcal{G}^{\underset{+-}{\mu\nu}}{}_{;\rho} = 0$ (i. e., such that $\Delta_\rho = f_\rho - f_{\bar{\bar{\rho}}}$).

4

The Second Group of Einstein's Equations

A. THE RIGOROUS FIELD EQUATIONS.

1. THE RIGOROUS FIELD EQUATIONS DEDUCED FROM $\mathcal{H}$.

If one starts from a variational principle applied to the density $\mathcal{H} = \mathcal{G}^{\mu\nu} R_{\mu\nu}$, $R_{\mu\nu}$ being Ricci's tensor $R_{\mu\nu}(\Gamma)$, one can obtain the second system (II) of the field equations. They are expressed in terms of an affine connection,

$$L^{\rho}_{\mu\nu} = \Gamma^{\rho}_{\mu\nu} + \frac{2}{3}\,\delta^{\rho}_{\mu}\,\Gamma_{\nu}, \quad L_{\rho} = L^{\sigma}_{\underset{\vee}{\rho\sigma}} = 0\,. \qquad (2.31)$$

Let us consider system II in which we shall include a cosmological term *

$$\left\{\begin{array}{ll} \text{a.} & D_{\rho}\,\mathcal{G}^{\underset{+-}{\mu\nu}} = 0, \quad L_{\rho} = 0 \text{ equivalent to } \mathfrak{g}^{\rho} = \partial_{\mu}\,\mathfrak{g}^{\rho\mu} = 0 \\ \text{b.} & \left\{\begin{array}{l} W_{\underline{\mu\nu}} = \lambda\gamma_{\mu\nu} \\ \partial_{\rho}\,W_{\underset{\vee}{\mu\nu}} + \partial_{\nu}\,W_{\underset{\vee}{\rho\mu}} + \partial_{\mu}\,W_{\underset{\vee}{\nu\rho}} = \lambda\,\varphi_{\mu\nu\rho}. \end{array}\right. \end{array}\right. \qquad \text{(II)}$$

---

* $R_{\mu\nu} = \lambda\,g_{\mu\nu}$ i.e. from (2.35)

$$W_{\underline{\mu\nu}} = \lambda\gamma_{\mu\nu} \qquad W_{\underset{\vee}{\mu\nu}} - \frac{2}{3}\left(\partial_{\mu}\,\Gamma_{\nu} - \partial_{\nu}\Gamma_{\mu}\right) = \lambda\varphi_{\mu\nu}$$

In II, we must express Ricci's tensor $W_{\mu\nu}(L_\rho)$ in terms of the fields and their derivatives by substituting the value of $L^{\rho}_{\mu\nu}$ obtained from (3.51) and (3.56).

Following the notations (3.55) adopted in the preceding chapter, we can write:

$$L^{\rho}_{\mu\nu} = \left\{ {\rho \atop \mu\nu} \right\} + u^{\rho}_{\mu\nu} + \underset{\vee}{L}^{\rho}_{\mu\nu} = \left\{ {\rho \atop \mu\nu} \right\} + \Theta^{\rho}_{\mu\nu} \tag{4.1}$$

where $\Theta^{\rho}_{\mu\nu} = u^{\rho}_{\mu\nu} + \underset{\vee}{L}^{\rho}_{\mu\nu}$ and the $\left\{ {\rho \atop \mu\nu} \right\}$ are the Christoffel symbols obtained from $\gamma_{\mu\nu}$.

Substitution of (4.1) in $W_{\mu\nu}$ as given in (2.36)

$$W_{\mu\nu} = \partial_\rho L^{\rho}_{\mu\nu} - \partial_\nu L^{\rho}_{\mu\rho} + L^{\lambda}_{\mu\nu} L^{\rho}_{\lambda\rho} - L^{\lambda}_{\mu\rho} L^{\rho}_{\lambda\nu}, \tag{2.36}$$

lead to:

$$W_{\mu\nu} = G_{\mu\nu} + \nabla_\rho \Theta^{\rho}_{\mu\nu} - \frac{1}{2} \nabla_\mu \nabla_\nu \operatorname{Log} g + \Theta^{\rho}_{\mu\nu} u_\rho - \Theta^{\lambda}_{\mu\rho} \Theta^{\rho}_{\lambda\nu}. \tag{4.2}$$

where $G_{\mu\nu}$ is the Ricci tensor formed with $\left\{ {\rho \atop \mu\nu} \right\}$, ∇_ρ is the covariant derivative with respect to $\left\{ {\rho \atop \mu\nu} \right\}$ and $u_\rho = u^{\mu}_{\rho\mu}$ (see 3.39). We can now separate $W_{\mu\nu}$ in terms of its symmetric and antisymmetric parts:

$$\begin{aligned} \underline{W_{\mu\nu}} = G_{\mu\nu} + \nabla_\rho u^{\rho}_{\mu\nu} - \frac{1}{2} \nabla_\mu \nabla_\nu \operatorname{Log} g + \\ + u^{\rho}_{\mu\nu} u_\rho - (\underline{u^{\lambda}_{\mu\rho}} u^{\rho}_{\lambda\nu} + \underset{\vee}{L}^{\lambda}_{\mu\rho} \underset{\vee}{L}^{\rho}_{\lambda\nu}). \end{aligned} \tag{4.3}$$

$$\underset{\vee}{W}_{\mu\nu} = \nabla_\mu \underset{\vee}{L}^{\rho}_{\mu\nu} + \underset{\vee}{L}^{\rho}_{\mu\nu} u_\rho - (u^{\lambda}_{\mu\rho} \underset{\vee}{L}^{\rho}_{\lambda\nu} + u^{\rho}_{\lambda\nu} \underset{\vee}{L}^{\lambda}_{\mu\rho}). \tag{4.4}$$

In the above relations, $u^{\rho}_{\mu\nu}$ is expressed in terms of $\underset{\vee}{L}^{\lambda}_{\sigma\tau}$. From (3.17), we have

$$\begin{aligned} u^{\rho}_{\mu\nu} &= -\gamma^{\rho\sigma}(\varphi_{\nu\lambda} \underset{\vee}{L}^{\lambda}_{\mu\sigma} + \varphi_{\mu\lambda} \underset{\vee}{L}^{\lambda}_{\nu\sigma}) \\ &= -\gamma^{\rho\sigma}(\underset{\vee}{L}_{\mu\sigma,\bar{\nu}} + \underset{\vee}{L}_{\nu\sigma,\bar{\mu}}). \end{aligned} \tag{3.17}$$

The field equations (II) can then be written in the form:

$$G_{\mu\nu} - \nabla^{\rho}(\underset{\vee}{L}_{\mu\rho,\bar{\nu}} + \underset{\vee}{L}_{\nu\rho,\bar{\mu}}) - \frac{1}{2}\nabla_{\mu}\nabla_{\nu} \log g$$
$$- \frac{1}{2}(\underset{\vee}{L}_{\mu\rho,\bar{\nu}} + \underset{\vee}{L}_{\nu\rho,\bar{\mu}})\partial^{\rho} \operatorname{Log}\frac{g}{\gamma} \tag{4.5}$$
$$+ \gamma^{\lambda\sigma}\gamma^{\rho\tau}[\underset{\vee}{L}_{\mu\rho,\sigma}\underset{\vee}{L}_{\nu\lambda,\tau} - (\underset{\vee}{L}_{\mu\sigma,\bar{\rho}} + \underset{\vee}{L}_{\rho\sigma,\bar{\mu}})$$
$$(L_{\nu\tau,\bar{\lambda}} + L_{\lambda\tau,\bar{\nu}})] = \lambda\gamma_{\mu\nu}$$

$$\nabla_{\sigma}\Big[\nabla_{\rho}\underset{\vee}{L}^{\rho}_{\mu\nu} + \frac{1}{2}\underset{\vee}{L}^{\rho}_{\mu\nu}\partial_{\rho}\operatorname{Log}\frac{g}{\gamma}$$
$$+ \gamma^{\lambda\tau}(2\underset{\vee}{L}^{\rho}_{\mu\nu}\underset{\vee}{L}_{\nu\sigma,\bar{\rho}} + \underset{\vee}{L}^{\rho}_{\mu\lambda}\underset{\vee}{L}_{\rho\tau,\bar{\nu}} - \underset{\vee}{L}^{\rho}_{\nu\lambda}\underset{\vee}{L}_{\rho\tau,\bar{\mu}})\Big] \tag{4.6}$$
$$+ \text{circular permutation of } \sigma,\ \mu,\ \nu = \lambda\,\varphi_{\mu\nu\rho}.$$

with

$$\nabla^{\rho} = \gamma^{\rho\sigma}\nabla_{\sigma},\quad \partial^{\rho} = \gamma^{\rho\sigma}\partial_{\sigma}. \tag{4.7}$$

In these expressions, one can substitute the solution of $\underset{\vee}{L}^{\rho}_{\mu\nu}$ obtained from (3. 51) and (3. 56).

In this form, the field equations (4. 5) and (4. 6) seem quite complicated. A resolution similar to the one obtained by Foures (118) for $G_{\mu\nu} = 0$ does not seem possible.

Insofar as the physical interpretation of these equations and their applications, one cannot consider more than the following:

1° One can keep the rigorous solutions (3. 51) -(3. 56) and apply them to special simple cases. An example of this is the search for a spherically symmetric solution (see Chapter 5).

2° One can seek an approximate form of the relations (4. 5) and (4. 6) (see § B this chapter).

2. DEFINITION OF THE CURRENT FOUR VECTOR.

The theory permits the a priori definition of two current densitites.

$$\mathfrak{J}^{\mu} = \partial_{\rho} \mathfrak{g}^{\mu\rho}, \tag{4.8}$$

$$\mathfrak{I}^{\mu} = \frac{1}{6\sqrt{-\gamma}} \epsilon^{\mu\nu\rho\sigma} \varphi_{\nu\rho\sigma} = \frac{1}{6\sqrt{-\gamma}} \epsilon^{\mu\nu\rho\sigma} (\partial_{\nu} \varphi_{\rho\sigma} + \partial_{\sigma}\varphi_{\nu\rho} + \partial_{\rho} \varphi_{\sigma\nu}). \tag{4.9}$$

But the current density $\mathfrak{J}^{\mu}$ satisfies the identity

$$\frac{1}{2} (\mathcal{G}^{\mu\rho}_{+-;\rho} - \mathcal{G}^{\rho\mu}_{+-;\rho}) \equiv \partial_{\rho} \mathfrak{g}^{\mu\rho} - \mathfrak{H}^{\mu\rho}\Gamma_{\rho}. \tag{4.51}$$

In particular, if this identity is expressed in terms of an affine connection Δ such that $\mathcal{G}^{\mu\nu}_{+-;\rho} = 0$, we always have:

$$\Delta_{\rho} = h_{\rho\sigma} f^{\sigma} \tag{2.40}$$

or

$$\Delta_{\rho} = f_{\rho} - f_{\bar{\bar{\rho}}}. \tag{2.41}$$

In the case where the variational principle is applied to a density formed by the Ricci tensor $R_{\mu\nu}(\Gamma)$, the relation $\mathcal{G}^{\mu\nu}_{+-;\rho} = 0$ is valid for the connection $L^{\rho}_{\mu\nu} = \Gamma^{\rho}_{\mu\nu} + 2/3\ \delta^{\rho}_{\mu}\Gamma_{\nu}$ whose torsion is zero. Thus $\Gamma_{\rho} = f_{\rho} - f_{\bar{\bar{\rho}}}$ is satisfied by $L_{\rho} = 0$, $f_{\rho} = 0$, these relations following from the variational principle. Thus, since $\mathfrak{J}^{\mu}$ is zero, we are led to believe that the existence of an electric current is necessarily tied to the definition of the pseudo vector $\mathfrak{I}^{\mu}$. If we assume for simplicity that the only non-zero components of the antisymmetric field* are φ_{23} and φ_{14} $\mathfrak{I}^{\mu}$, reduces to in polar coordinates

* This is the case for the spherically symmetric static field.

$$\mathcal{J}^4 = \frac{-1}{\sqrt{-\gamma}} \partial_1 \varphi_{23} .$$

We are then led to identify φ_{23} with the electric field and φ_{24} with the magnetic field. This identification is different from the usual one (in Maxwell's theory, φ_{p4} is identified with the electric field and φ_{pq} (p, q = 1, 2, 3) with the magnetic field). We shall see however that this identification is better from the point of view of the unified theory.

In fact, Einstein noted (7) that it is possible and probably preferable to identify the electric field with the components φ_{pq}(p, q = 1, 2, 3) of an antisymmetric tensor and the magnetic field with the components φ_{p4}. In this manner, the sign of the magnetic field and the current automatically change sign when one reverses the direction of time. In discussing this question Schrödinger (75, 76) points out that the disymmetry between the electric and magnetic cases which manifests itself in unified theory is one of the most satisfying characters of the unified theory. This disymmetry manifests itself in the particular case of spherically symmetric solutions.

Nevertheless, it does not seem evident to us that the only possible definition of the electric current is (4.9). We shall give in Chapter 6 another possible definition from which it will not be evident that the $\varphi_{\rho q}$ represent the electric field. The latter can be identified with φ_{p4}.*

B. APPROXIMATE EQUATIONS.

3. DERIVATION OF THE APPROXIMATE EQUATIONS.

If the fields $\varphi_{\mu\nu}$ are weak and if the $\gamma_{\mu\nu}$ are close to the

---

* In this case, a spherically symmetric solution for such a Papapetrou solution can still be associated with the electric field case when one chooses the equations of the weak system. The corresponding equations for the strong system cannot represent the electric field case (cf. Chapter 5, p. 86 and (61), (63)). But the natural equations here are the equations of the weak system.

flat space values $\eta_{\mu\nu}(-1,\ -1,\ -1,\ +1)$, we can expand the $g_{\mu\nu}$ in powers of ϵ:

$$\gamma_{\mu\nu} = \eta_{\mu\nu} + \epsilon \underset{1}{\gamma}_{\mu\tau} + \epsilon^2 \underset{2}{\gamma}_{\mu\tau} + \epsilon^3 \underset{3}{\gamma}_{\mu\nu} + \ldots, \tag{4.10}$$

$$\varphi_{\mu\nu} = \epsilon \underset{1}{\varphi}_{\mu\nu} + \epsilon^2 \underset{2}{\varphi}_{\mu\nu} + \epsilon^3 \underset{3}{\varphi}_{\mu\nu} + \ldots \tag{4.11}$$

The affine connection then becomes:

$$\left\{ {\rho \atop \mu\nu} \right\} = \epsilon \underset{1}{\left\{ {\rho \atop \mu\nu} \right\}} + \epsilon^2 \underset{2}{\left\{ {\rho \atop \mu\nu} \right\}} + \epsilon^3 \underset{3}{\left\{ {\rho \atop \mu\nu} \right\}} + \ldots, \tag{4.12}$$

$$\underset{\vee}{L}{}^{\rho}_{\mu\nu} = \epsilon \underset{\vee}{\underset{1}{L}}{}^{\rho}_{\mu\nu} + \epsilon^2 \underset{\vee}{\underset{2}{L}}{}^{\rho}_{\mu\nu} + \epsilon^3 \underset{\vee}{\underset{3}{L}}{}^{\rho}_{\mu\nu} + \ldots \tag{4.13}$$

and from (3. 17):

$$\begin{aligned} u^{\rho}_{\mu\nu} &= -\gamma^{\rho\sigma}(\underset{\vee}{L}_{\mu\sigma,\bar{\nu}} + \underset{\vee}{L}_{\nu\sigma,\bar{\mu}}) \\ &= -\gamma^{\rho\sigma}(\varphi_{\nu\lambda}\underset{\vee}{L}{}^{\lambda}_{\mu\sigma} + \varphi_{\mu\lambda}\underset{\vee}{L}{}^{\lambda}_{\nu\sigma}) \\ &= -\epsilon^2 \underset{2}{u}{}^{\rho}_{\mu\nu} + \epsilon^3 \underset{3}{u}{}^{\rho}_{\mu\nu} + \ldots \end{aligned} \tag{4.14}$$

The field equations

$$\partial_{\rho}(\sqrt{-g}\, f^{\mu\rho}) = 0, \tag{4.15}$$

$$\partial_{\rho} \underset{\vee}{W}_{\mu\nu} + \partial_{\nu} \underset{\vee}{W}_{\rho\mu} + \partial_{\mu} \underset{\vee}{W}_{\nu\rho} = 0, \tag{4.16}$$

$$W_{\underline{\mu\nu}} = 0 \tag{4.17}$$

can then be developed to different orders of approximations. From relations (1.18a), (4.3) and (4.4) for $f^{\mu\rho}$, $W_{\underline{\mu\nu}}$ and $\underset{\vee}{W}_{\mu\nu}$, we then have the following.

First order:

$$\eta^{\rho\nu}\partial_\rho \underset{1}{\varphi}_{\mu\nu} = 0, \tag{4.18}$$

$$\eta^{\rho\nu}\partial_\tau[\partial_\rho \underset{\substack{1\\V}}{L}_{\mu\sigma,\,\nu}] + \text{circular permutation of } \tau,\ \mu,\ \sigma = 0, \tag{4.19}$$

$$\underset{1}{G}_{\mu\nu} = \underset{1}{\mathcal{L}}_{\mu\nu} = 0, \tag{4.20}$$

Let

$$\underset{\substack{1\\V}}{L}{}^{\rho}_{\mu\nu} = \eta^{\rho\sigma}\underset{\substack{1\\V}}{L}_{\mu\nu,\,\sigma}, \tag{4.21}$$

$$\mathcal{L}_{\mu\nu}(\gamma) = \frac{1}{2}\eta^{\rho\sigma}(\partial_\rho\partial_\mu\gamma_{\nu\sigma} - \partial_\rho\partial_\sigma\gamma_{\mu\nu} - \partial_\nu\partial_\mu\gamma_{\rho\sigma} + \partial_\nu\partial_\sigma\gamma_{\mu\sigma}) \tag{4.22}$$

and

$$\underset{1}{\mathcal{L}}_{\mu\nu} = \mathcal{L}_{\mu\nu}(\underset{1}{\gamma}), \qquad \underset{2}{\mathcal{L}}_{\mu\nu} = \mathcal{L}_{\mu\nu}(\underset{2}{\gamma}). \tag{4.23}$$

To second order:

$$\eta^{\rho\nu}\partial_\rho\underset{2}{\varphi}_{\mu\nu} + \eta^{\rho\nu}\underset{1}{\varphi}_{\mu\nu}(\partial_\rho\sqrt{-g})_1 - \eta^{\sigma\lambda}\eta^{\rho\nu}\underset{1}{\varphi}_{\sigma\nu}\partial_\rho\underset{1}{\gamma}_{\mu\nu} - \eta^{\rho\nu}\eta^{\sigma\lambda}\partial_\rho(\underset{1}{\gamma}_{\nu\lambda}\underset{1}{\varphi}_{\mu\sigma}) = 0, \tag{4.24}$$

$$\partial_\tau\left[\partial_\rho\underset{\substack{2\\V}}{L}{}^{\rho}_{\mu\nu} + \underset{1}{\left\{\begin{matrix}\rho\\ \sigma\rho\end{matrix}\right\}}\underset{\substack{1\\V}}{L}{}^{\sigma}_{\mu\nu} - \underset{1}{\left\{\begin{matrix}\lambda\\ \mu\rho\end{matrix}\right\}}\underset{\substack{1\\V}}{L}{}^{\rho}_{\lambda\nu} - \underset{1}{\left\{\begin{matrix}\lambda\\ \nu\rho\end{matrix}\right\}}\underset{\substack{1\\V}}{L}{}^{\rho}_{\mu\nu}\right] + \text{circular permutation of } \tau,\ \mu,\ \nu = 0, \tag{4.25}$$

$$\underset{2}{G}_{\mu\nu} + \partial_\rho\underset{2}{u}{}^{\rho}_{\mu\nu} - \frac{1}{2}\partial_\mu\partial_\nu \operatorname{Log}\left(\frac{g}{\gamma}\right)_2 - \underset{\substack{1\\V}}{L}{}^{\lambda}_{\mu\rho}\underset{\substack{1\\V}}{L}{}^{\rho}_{\lambda\nu} = 0. \tag{4.26}$$

On the other hand, the general solution (3.51) - (3.56) allows us to compute the affine connection to any order. We limit ourselves to the third order. The coefficients a and b defined in (3. 52) can be written:

$$a = 2 - \left(\frac{g}{\gamma}\right)_2 - \left(\frac{g}{\gamma}\right)_3 + \ldots, \qquad b = \frac{6(\sqrt{\varphi})_2}{\sqrt{-\gamma}} . \tag{4. 27}$$

The relations (3. 54) applied to $L_{\underset{\vee}{\mu\nu},\rho} = \gamma_{\rho\sigma} L^{\sigma}_{\underset{\vee}{\mu\nu}}$ can be written

$$4\left[1 - \left(\frac{g}{\gamma}\right)_2 - \left(\frac{g}{\gamma}\right)_3 + \ldots\right]\left(\underset{1}{L}_{\underset{\vee}{\mu\nu},\rho} + \underset{2}{L}_{\underset{\vee}{\mu\nu},\rho} + \underset{3}{L}_{\underset{\vee}{\mu\nu},\rho} + \ldots\right)$$

$$= \left[2 - \left(\frac{g}{\gamma}\right)_2 - \left(\frac{g}{\gamma}\right)_3\right]\left(\underset{1}{S}_{\underset{\vee}{\mu\nu},\rho} + \underset{2}{S}_{\underset{\vee}{\mu\nu},\rho} + \underset{3}{S}_{\underset{\vee}{\mu\nu},\rho} + \ldots\right) \tag{4. 28}$$

$$+ 6(\sqrt{\varphi})_2 \underset{1}{S}^{*}_{\underset{\vee}{\mu\nu},\rho},$$

that is

$$2\underset{1}{L}_{\underset{\vee}{\mu\nu},\rho} = \underset{1}{S}_{\mu\nu,\rho}, \tag{4. 29}$$

$$2\underset{2}{L}_{\underset{\vee}{\mu\nu},\rho} = \underset{2}{S}_{\underset{\vee}{\mu\nu},\rho}, \tag{4. 30}$$

$$2\underset{3}{L}_{\underset{\vee}{\mu\nu},\rho} - 2\left(\frac{g}{\gamma}\right)_2 \underset{1}{L}_{\underset{\vee}{\mu\nu},\rho} = \underset{3}{S}_{\underset{\vee}{\mu\nu},\rho} - \frac{1}{2}\left(\frac{g}{\gamma}\right)_2 \underset{1}{S}_{\underset{\vee}{\mu\nu},\rho}$$
$$+ 3\,(\sqrt{\varphi})_2\, \underset{1}{S}^{*}_{\underset{\vee}{\mu\nu},\rho}, \tag{4. 31}$$

or

$$2\underset{3}{L}_{\underset{\vee}{\mu\nu},\rho} = \underset{3}{S}_{\underset{\vee}{\mu\nu},\rho} + \frac{1}{2}\left(\frac{g}{\gamma}\right)_2 \underset{1}{S}_{\underset{\vee}{\mu\nu},\rho} + 3(\sqrt{\varphi})_2\, \underset{1}{S}^{*}_{\underset{\vee}{\mu\nu},\rho} . \tag{4. 32}$$

But from (3. 53)

$$\underset{1}{S}_{\underset{\vee}{\mu\nu},\rho} + \underset{2}{S}_{\underset{\vee}{\mu\nu},\rho} + \underset{3}{S}_{\underset{\vee}{\mu\nu},\rho} + \ldots$$

$$= \left[2 - \left(\frac{g}{\gamma}\right)_2 - \left(\frac{g}{\gamma}\right)_3\right]\left(\underset{1}{R}_{\underset{\vee}{\mu\nu},\rho} + \underset{2}{R}_{\underset{\vee}{\mu\nu},\rho} + \underset{3}{R}_{\underset{\vee}{\mu\nu},\rho}\right) \tag{4.33}$$

$$- 2(\sqrt{\varphi})_2\, \underset{1}{R}^{*}_{\underset{\vee}{\mu\nu},\rho} - \underset{3}{R}_{\underset{\vee}{\mu\nu},\bar{\bar{\rho}}};$$

Hence

$$\underset{1}{S}_{\underset{V}{\mu\nu},\rho} = 2\,\underset{1}{R}_{\underset{V}{\mu\nu},\rho}, \tag{4.34}$$

$$\underset{2}{S}_{\underset{V}{\mu\nu},\rho} = 2\,\underset{2}{R}_{\underset{V}{\mu\nu},\rho}, \tag{4.35}$$

$$\underset{3}{S}_{\underset{V}{\mu\nu},\rho} = 2\,\underset{3}{R}_{\underset{V}{\mu\nu},\rho} - \left(\frac{g}{\gamma}\right)_2 \underset{1}{R}_{\underset{V}{\mu\nu},\rho} - 2(\sqrt{\varphi})_2\,\underset{1}{R}_{\underset{V}{\mu\nu},\rho} - \underset{3}{R}_{\underset{V}{\mu\nu},\bar{\bar{\rho}}} \tag{4.36}$$

and, comparing (4.29)-(4.32), we have:

$$\underset{1}{L}_{\underset{V}{\mu\nu},\rho} = \underset{1}{R}_{\underset{V}{\mu\nu},\rho}, \tag{4.37}$$

$$\underset{2}{L}_{\underset{V}{\mu\nu},\rho} = \underset{2}{R}_{\underset{V}{\mu\nu},\rho}, \tag{4.38}$$

$$\underset{3}{L}_{\underset{V}{\mu\nu},\rho} = \underset{3}{R}_{\underset{V}{\mu\nu},\rho} + 2(\sqrt{\varphi})_2\,\underset{1}{\overset{*}{R}}_{\underset{V}{\mu\nu},\rho} - \frac{1}{2}\,\underset{3}{R}_{\underset{V}{\mu\nu},\bar{\bar{\rho}}} \tag{4.39}$$

Use of (3.56) leads to:

$$\underset{1}{L}_{\underset{V}{\mu\nu},\rho} = \partial_\rho\,\underset{1}{\varphi}_{\mu\nu} - \frac{1}{2}\,\underset{1}{\varphi}_{\mu\nu\rho}, \tag{4.40}$$

$$\underset{2}{L}_{\underset{V}{\mu\nu},\rho} = \partial_\rho\,\underset{2}{\varphi}_{\mu\nu} - \frac{1}{2}\,\underset{2}{\varphi}_{\mu\nu\rho} - \underset{1}{\left\{\begin{matrix}\sigma\\ \mu\rho\end{matrix}\right\}}\underset{1}{\varphi}_{\sigma\nu} - \underset{1}{\left\{\begin{matrix}\sigma\\ \nu\rho\end{matrix}\right\}}\underset{1}{\varphi}_{\mu\sigma}, \tag{4.41}$$

$$\begin{aligned}\underset{3}{L}_{\underset{V}{\mu\nu},\rho} = {} & \partial_\rho\,\underset{3}{\varphi}_{\mu\nu} - \frac{1}{2}\,\underset{3}{\varphi}_{\mu\nu\rho} - \underset{1}{\left\{\begin{matrix}\sigma\\ \mu\rho\end{matrix}\right\}}\underset{2}{\varphi}_{\sigma\nu} - \underset{1}{\left\{\begin{matrix}\sigma\\ \nu\rho\end{matrix}\right\}}\underset{2}{\varphi}_{\mu\sigma} - \underset{2}{\left\{\begin{matrix}\sigma\\ \mu\rho\end{matrix}\right\}}\underset{1}{\varphi}_{\sigma\nu} \\ & - \underset{2}{\left\{\begin{matrix}\sigma\\ \nu\rho\end{matrix}\right\}}\underset{1}{\varphi}_{\mu\sigma} - \frac{1}{2}(\sqrt{\varphi})_2\,\overset{*}{\varphi}_{(\mu\nu)\rho} - \underset{1}{\varphi}_{\mu\nu}\,\partial_\rho \operatorname{Log}\left(\frac{g}{\gamma}\right)_2 \\ & + \overset{*}{\epsilon}_{|\mu\nu|\rho\lambda}(\sqrt{\varphi}\,\varphi^{\sigma\lambda})_1\,\partial_\sigma(\sqrt{\varphi})_2 - \underset{1}{\overset{*}{\varphi}}_{\mu\nu}\,\partial_\rho(\sqrt{\varphi})_2 \\ & + \frac{1}{4}\,\underset{1}{\overset{*}{\varphi}}_{\mu\nu}(\sqrt{\varphi}\,\varphi^{\lambda\tau})_1\,\underset{1}{\varphi}_{\lambda\tau\rho} + \frac{1}{2}\,\epsilon_{\mu\nu\rho\lambda}(\sqrt{\varphi}\,\varphi^{\sigma\lambda})_1\,\partial_\sigma \operatorname{Log}\left(\frac{g}{\gamma}\right)_2 \\ & + 2(\sqrt{\varphi})_2\,\partial_\rho\,\underset{1}{\overset{*}{\varphi}}_{\mu\nu} - \frac{1}{2}\,\partial_{\bar{\bar{\rho}}}\,\underset{1}{\varphi}_{\mu\nu} + \frac{1}{4}\,\underset{3}{\varphi}_{\mu\nu\bar{\bar{\rho}}}.\end{aligned} \tag{4.42}$$

It is sufficient to replace $\underset{V}{{}_1L}_{\mu\nu,\rho}$, $\underset{V}{{}_2L}_{\mu\nu,\rho}$, $\underset{V}{{}_3L}_{\mu\nu,\rho}$ (and consequently $\underset{2}{u}_{\mu\nu,\rho}$ and $\underset{3}{u}_{\mu\nu,\rho}$ from (4. 14)) by their values (4. 40) - (4. 42) in the various equations to have the approximate relations to third order.

The development of the relations have been carried to second order by various authors. Three different forms have been given.

4. THE APPROXIMATE EQUATIONS ($\underset{1}{\gamma}_{\mu\nu} \neq 0$).

The development of the field equations to second order was carried out by Einstein and Kaufman(18).

a. Equations of the symmetrical field. Substitution of (4. 40) and (4. 14) in (4. 26) leads to:

$$\underset{1}{\mathcal{L}}_{\mu\nu} = 0, \tag{4. 43}$$

$$\underset{2}{\mathcal{L}}_{\mu\nu} + \underset{2}{Q}_{\mu\nu} = 0, \tag{4. 44}$$

$\underset{1}{\mathcal{L}}_{\mu\nu}$ and $\underset{2}{\mathcal{L}}_{\mu\nu}$ being defined by (4. 22) and (4. 23). Here $Q_{\mu\nu}$ has the form

$$\begin{aligned} Q_{\mu\nu} = \eta^{\rho\lambda}\eta^{\sigma\tau}\Big\{ & \tfrac{1}{2}(\partial_\mu\partial_\nu\gamma_{\rho\sigma} + \partial_\rho\partial_\sigma\gamma_{\mu\nu} - \partial_\rho\partial_\mu\gamma_{\nu\sigma} - \partial_\nu\partial_\sigma\gamma_{\mu\rho})\gamma_{\lambda\tau} \\ & - [\mu\nu,\sigma]\partial_\rho\gamma_{\lambda\tau} + [\mu\rho,\sigma]\partial_\nu\gamma_{\lambda\tau} + [\sigma\rho,\lambda][\mu\nu,\tau] \\ & - [\sigma\nu,\lambda][\mu\rho,\tau] - \tfrac{1}{4}\partial_\mu\partial_\nu\varphi_{\rho\sigma}\varphi_{\lambda\tau} + \partial_\tau\varphi_{\lambda\nu}\partial_\rho\varphi_{\mu\sigma} \\ & - \tfrac{1}{4}\varphi_{\sigma\nu\lambda}\,\varphi_{\tau\mu\rho} + \tfrac{1}{2}(\varphi_{\nu\sigma}\partial_\rho\,\varphi_{\tau\mu\lambda} + \varphi_{\mu\sigma}\partial_\rho\,\varphi_{\tau\nu\lambda})\Big\}. \end{aligned} \tag{4. 45}$$

Einstein and Kaufman have shown that to each solution of (4. 43) corresponds a solution of the second order equations (4. 44).

b. Equations of the antisymmetrical field. These are equations (4. 18), (4. 19), (4. 24) and (4. 25). The first order equations (4. 18) and (4. 19) will be:

$$\eta^{\rho\nu}\,\partial_\rho \underset{1}{\varphi}_{\mu\nu} = 0, \tag{4.18}$$

$$\eta^{\rho\nu}\partial_\rho\partial_\nu \underset{1}{\varphi}_{\mu\tau\lambda} = 0. \tag{4.46}$$

The second order is obtained by substituting (4. 40) into (4. 19) and taking (4. 18) into account. They can then be written in the form

$$\eta^{\rho\nu}[\partial_\rho \underset{2}{\varphi}_{\mu\nu} \quad \underset{1}{\varphi}_{\mu\nu}(\partial_\rho\sqrt{-g})_1 - \eta^{\sigma\lambda}\partial_\rho(\underset{1}{\gamma}_{\nu\lambda}\underset{1}{\varphi}_{\mu\sigma})$$
$$- \eta^{\sigma\lambda}\underset{1}{\varphi}_{\sigma\nu}\partial_\rho \underset{1}{\gamma}_{\mu\lambda}] = 0, \tag{4.47}$$

$$\partial_\epsilon \eta^{\rho\sigma}[\partial_\rho \underset{2}{\overset{}{L}}_{\underset{V}{\mu\nu},\sigma} - \eta^{\lambda\tau}(\underset{1}{\gamma}_{\sigma\tau}\partial_\rho \underset{1}{L}_{\underset{V}{\mu\nu},\lambda} + \underset{1}{L}_{\underset{V}{\mu\nu},\lambda}\partial_\rho \underset{1}{\gamma}_{\sigma\tau}$$
$$- [\underset{1}{\lambda\rho}, \sigma]\,\underset{1}{L}_{\underset{V}{\mu\nu},\tau} + [\underset{1}{\mu\rho}, \tau]\underset{1}{L}_{\underset{V}{\lambda\nu},\sigma}$$
$$+ [\nu\rho, \tau]\underset{1}{L}_{\underset{V}{\mu\lambda},\sigma})] \tag{4.48}$$
$$+ \text{circular permutation of } \mu, \nu, \epsilon = 0.$$

The significance of these equations was discussed by Einstein and Kaufman in the case of the strong system, that is, when the circular permutation in (4. 19), (4. 25) and (4. 48) is replaced by that quantity which is controlled by ∂_τ (or ∂_ϵ). The physical consequences of the conclusions reached seem to exclude the strong system. In fact, in this case, the first order equations (4. 18) and (4. 19) reduce to

$$\eta^{\rho\nu}\partial_\rho \underset{1}{\varphi}_{\mu\nu} = 0, \tag{4.18}$$

$$\gamma^{\rho\sigma}\partial_\rho\partial_\sigma \underset{1}{\varphi}_{\mu\nu} = 0, \tag{4.49}$$

since

$$\eta^{\rho\sigma}\partial_\rho \underset{1}{\varphi}_{\mu\nu\sigma} = 0.$$

In order to insure the existence of solutions of the second order approximation, Einstein and Kaufman have shown that the condition

$$\partial_\rho \underset{1}{\varphi}_{\mu\nu}(\partial_\rho \partial_\nu \underset{1}{\gamma}_{\sigma\mu} + \partial_\sigma \partial_\mu \underset{1}{\gamma}_{\rho\nu}) = 0 \qquad (4.50)$$

must always be satisfied. To first order, $\underset{1}{\gamma}_{\mu\nu}$ and $\underset{1}{\varphi}_{\mu\nu}$ satisfy (4.18) and (4.20) and are independent of each other. We could then arbitrarily superpose the two fields. This condition is also applicable to the weak system. But to second order (4.50) leads to an important restriction to the superposition of these fields. One can show that in the case of a Schwarzschild field that (4.50) will impose restrictions which are physically inacceptable ((18) p. 334). The additive property of the weak fields cannot be restricted by the stringent condition (4.50). For this strictly physical reason, the strong system must be rejected.

Schrödinger (20) has shown that the equations of the weak system lead to the following conditions for the antisymmetrical field in second order:

$$\eta^{\mu\nu}\partial_\nu \underset{2}{\varphi}_{\mu\rho} + \underset{2}{p}_\rho = 0, \qquad (4.51)$$

$$\frac{1}{2}\underset{2}{\varphi}_{\mu\nu\rho} = \underset{2}{V}_{\mu\nu\rho}, \qquad (4.52)$$

where

$$\underset{2}{\varphi}_{\mu\nu\rho} = \partial_\mu \underset{2}{\varphi}_{\nu\rho} + \partial_\rho \underset{2}{\varphi}_{\mu\nu} + \partial_\nu \underset{2}{\varphi}_{\rho\mu}, \qquad (4.53)$$

$$\underset{2}{V}_{\mu\nu\rho} = \partial_\mu \underset{2}{V}_{\nu\rho} + \partial_\rho \underset{2}{V}_{\mu\nu} + \partial_\nu \underset{2}{V}_{\rho\mu} \qquad (4.54)$$

p_ρ and $V_{\mu\nu}$ being quadratic expressions.
We then have identically

$$\eta^{\rho\sigma}\partial_\sigma \underset{2}{p}_\rho \equiv 0 \qquad (4.55)$$

On the other hand, if we choose a coordinate system such that

$$\eta^{\rho\sigma}(\partial_\rho \underset{1}{\gamma}_{\mu\sigma} - \frac{1}{2}\partial_\mu \underset{1}{\gamma}_{\rho\sigma}) = 0 \qquad (4.56)$$

we can reduce (4.52) to

$$\frac{1}{6}\epsilon^{\mu\nu\rho\sigma}\eta^{\lambda\tau}\partial_\lambda \partial_\tau \underset{2}{\varphi}_{\mu\nu\rho} = \epsilon^{\mu\nu\rho\sigma}\partial_\rho \underset{2}{V}_{\mu\nu}, \qquad (4.57)$$

or

$$\frac{1}{6}\,\epsilon^{\mu\nu\rho\sigma}\eta^{\lambda\tau}\partial_\lambda\partial_\tau \underset{2}{\varphi}_{\mu\nu\rho} = -\epsilon^{\mu\nu\rho\sigma}\eta^{\lambda\tau}\eta^{\epsilon\delta}(\partial_\nu \underset{1}{\varphi}_{\tau\delta})\, \underset{1}{G}_{\lambda\epsilon\rho\mu}, \tag{4.58}$$

where G is the curvature tensor formed by the γ to first order.* The right-hand side of (4. 58) does not vanish. Thus, even to the second order, the condition

$$\underset{2}{\varphi}_{\mu\nu\rho} = 0$$

cannot be realized. On the other hand, the divergence of the right hand side of (4. 58) does vanish and leads to the existence of solutions

$$\epsilon^{\mu\nu\rho\sigma}\,\partial_\sigma \underset{2}{\varphi}_{\mu\nu\rho} = 0 \tag{4.59}$$

that satisfy the continuity equation.

The order of magnitude of $\underset{2}{\varphi}_{\mu\nu\rho}$ could be determined from that of the Riemannian curvature if one knew the value of $\underset{1}{\varphi}_{\mu\nu}$. Unfortunately, we do not even know the units in which the electromagnetic field is expressed in terms of $\varphi_{\mu\nu}$.

Schrödinger has applied (4. 58) to the static fields. He shows that a purely electric or magnetic field in the first order remains unchanged in the second order. An application to the terrestrial field leads to the existence of a second order field which is α/R times the first order, being the gravitational radius of the earth. But the first order field D/R^3 is of the order of a gauss, D being the magnetic dipole moment of the earth. It is thus not probable that one can demonstrate experimentally the existence of this field and the current $\underset{2}{\varphi}_{\mu\nu\rho}$. If the relationship between the current and the two types of field is of theoretical importance, it nevertheless is outside the realm of experimental measurement.

---

* The difference in sign with Schrödinger's (26) Eq. 27 arises in the definition of the Riemann-Christoffel tensor.

5. THE APPROXIMATE EQUATIONS $\underset{1}{\gamma}_{\mu\nu} = 0$.

If we assume that the $\gamma_{\mu\nu}$ differs from the $\eta_{\mu\nu}$ in the second order only (i. e., $\underset{1}{\gamma}_{\mu\nu}$), the second order equations simplify considerably. Schrödinger (117) has discussed them for the strong system. We shall derive them for the weak system.

a. First order equations: $\underset{1}{L}_{\mu\nu}$ vanishes and the first order equations refer only to the antisymmetric field. From (4. 18) and (4. 19) we have:

$$\eta^{\rho\nu}\partial_\rho \underset{1}{\varphi}_{\mu\nu} = 0, \tag{4. 18}$$

$$\eta^{\rho\sigma}\partial_\rho\partial_\sigma \underset{1}{\varphi}_{\mu\nu\lambda} = 0. \tag{4. 46}$$

Using Schrödinger notation:

$$\eta_{\mu\nu} = \epsilon_\mu \; \delta_{\mu\nu}, \quad \text{for } \epsilon_\mu = -1, \; -1, \; -1, \; +1 \tag{4. 60}$$

where we do not sum over μ, (4. 18) and (4. 46) become

$$\epsilon_\rho \partial_\rho \underset{1}{\varphi}_{\mu\rho} = 0, \tag{4. 61}$$

$$\epsilon_\rho \partial_\rho \partial_\rho \underset{1}{\varphi}_{\mu\nu\lambda} = 0. \tag{4. 62}$$

b. Second order equations: These reduce simply to (4. 44):

$$\underset{2}{\mathcal{L}}_{\mu\nu} + \underset{2}{Q}_{\mu\nu} = 0 \tag{4. 44}$$

From (4. 45) and $\underset{1}{\gamma}_{\mu\nu} = 0$, $\underset{2}{Q}_{\mu\nu}$ reduces to

$$\underset{2}{Q}_{\mu\nu} = -\eta^{\rho\lambda}\eta^{\sigma\tau} \left\{ \frac{1}{4} \partial_\mu \partial_\nu \underset{1}{\varphi}_{\rho\sigma} \underset{1}{\varphi}_{\lambda\tau} - \partial_\tau \underset{1}{\varphi}_{\lambda\nu} \partial_\rho \underset{1}{\varphi}_{\mu\sigma} \right.$$
$$+ \frac{1}{4} \underset{1}{\varphi}_{\sigma\nu\lambda} \underset{1}{\varphi}_{\tau\mu\rho} - \frac{1}{2} \underset{1}{\varphi}_{\nu\sigma} \partial_\rho \underset{1}{\varphi}_{\tau\mu\lambda} \tag{4. 63}$$
$$\left. - \frac{1}{2} \underset{1}{\varphi}_{\mu\sigma} \partial_\rho \underset{1}{\varphi}_{\tau\nu\lambda} \right\}.$$

We then have, in Schrödinger's notation:

$$\underset{2}{\mathcal{L}}_{\mu\nu} = \epsilon_\rho \epsilon_\sigma \left\{ \frac{1}{4} \partial_\mu \partial_\nu \underset{1}{\varphi}_{\rho\sigma} \underset{1}{\varphi}_{\rho\sigma} - \partial_\sigma \underset{1}{\varphi}_{\rho\nu} \partial_\rho \underset{1}{\varphi}_{\mu\sigma} \right.$$
$$+ \frac{1}{4} \underset{1}{\varphi}_{\sigma\nu\rho} \underset{1}{\varphi}_{\sigma\mu\rho} - \frac{1}{2} \underset{1}{\varphi}_{\nu\sigma} \partial_\rho \underset{1}{\varphi}_{\sigma\mu\rho} \qquad (4.64)$$
$$\left. - \frac{1}{2} \underset{1}{\varphi}_{\mu\sigma} \partial_\rho \underset{1}{\varphi}_{\sigma\nu\rho} \right\}.$$

We will discuss this relation in the study of the energy tensor.*

---

* In the strong system, (4.18) is not changed, (4.19) must be replaced by

$$\underset{V}{W}_{\mu\nu} = 0 \quad \text{or} \quad \eta^{\rho\nu} \partial_\rho \underset{1}{\underset{V}{L}}_{\mu\sigma,\nu} = 0. \qquad (4.19')$$

We then have for the antisymmetric field

$$\eta^{\rho\nu} \partial_\rho \underset{1}{\varphi}_{\mu\nu} = 0, \qquad (4.18)$$

$$\eta^{\rho\sigma} \partial_\rho \partial_\sigma \underset{1}{\varphi}_{\mu\nu} = 0. \qquad (4.46')$$

From the definition of $\varphi_{\mu\nu\rho}$, we have

$$\eta^{\rho\sigma} \partial_\sigma \underset{1}{\varphi}_{\mu\nu\rho} = 0, \qquad (4.46'')$$

that is, if we use the notation (4.60)

$$\epsilon_\rho \partial_\rho \underset{1}{\varphi}_{\mu\rho} = 0. \qquad \epsilon_\rho \partial_\rho \underset{1}{\varphi}_{\mu\nu\rho} = 0. \qquad (4.61')$$

$\underset{2}{\mathcal{L}}_{\mu\nu}$ is then the expression (4.23) which Schrödinger denotes by $\Phi_{\mu\nu}$. In the strong system, (4.61') is valid and $\underset{2}{\mathcal{L}}_{\mu\nu}$ reduces to

$$\Phi_{\mu\nu} = \epsilon_\rho \epsilon_\sigma \left\{ \frac{1}{4} \partial_\mu \partial_\nu \underset{1}{\varphi}_{\rho\sigma} + \partial_\sigma \underset{1}{\varphi}_{\rho\nu} \partial_\rho \underset{1}{\varphi}_{\sigma\mu} + \frac{1}{4} \underset{1}{\varphi}_{\sigma\nu\rho} \underset{1}{\varphi}_{\sigma\mu\rho} \right\}.$$

which is the same as Schrödinger's expression (cf.(117), p. 21 Eq. (2.17)).

6. QUASI-STATIC SOLUTIONS.

We can assume that, in certain problems, the time derivatives will be of a lower order of magnitude than the space derivatives. We must then distinguish between g_{pq} and g_{p4} ($p = 1, 2, 3$) in an expansion in terms of ϵ. These solutions have been studied by Einstein and Infeld (94). Callaway (87) has applied this study to the weak system. We shall discuss his results in Chapter 7.

5

Spherically Symmetric Solution

A. DIFFERENTIAL EQUATIONS OF THE SPHERICALLY SYMMETRIC CASE.

1. SPHERICALLY SYMMETRIC CASE IN GENERAL RELIABILITY.

We shall seek the solutions $g_{\mu\nu}$ of Einstein's equations (II) for the case where the source possesses spherical symmetry.

a. Neutral particle. In general Relativity the problems is stated as follows: the field equations in the neighborhood of attractive masses

$$G_{\mu\nu} = 0 \tag{5. 1}$$

can be written as a function of the Christoffel symbols which are determined in the terms of the metric $\mathring{a}_{\mu\nu}$. We thus have a system of ten differential equations which satisfy four identities. When the given masses are spherically symmetric, it is convenient to work in polar coordinates. Then the line element is

$$ds^2 = -\mathring{\alpha}\, dr^2 - \mathring{\beta}(d\theta^2 + \sin^2\theta\, d\varphi^2) + \mathring{\sigma}\, dt^2 \tag{5. 2}$$

where the independent variables are:

$$x^1 = r, \quad x^2 = 0, \quad x^3 = \varphi, \quad x^4 = t. \tag{5. 3}$$

$\mathring{\alpha}$, $\mathring{\beta}$ and $\mathring{\sigma}$ depend on r and eventually on t. Computation of Ricci's tensor in terms of the $\mathring{a}_{\mu\nu}(-\mathring{\alpha}, -\mathring{\beta}, -\mathring{\beta}\sin^2\theta, \mathring{\sigma})$ leads to three differential equations. These can be reduced to two owing to the identity relations (cf. (4) p. 000 and (5) p. 85). We can then choose arbitrarily one of the unknowns $\mathring{\alpha}$, $\mathring{\beta}$, $\mathring{\sigma}$. It is convenient to choose $\mathring{\beta} = r^2$. Then the solutions in the static case are

$$\mathring{\beta} = r^2, \quad \mathring{\sigma} = \frac{V^2}{\mathring{\alpha}} = V^2 - \frac{2m}{r}, \tag{5.4}$$

V being the speed of light which we shall set equal to unity.

This is the Schwarzschild solution of the field equations. It defines completely the gravitational field in the neighborhood of attractive masses and permits the determination of the trajectories of particles moving in it. Although the field is computed for a purely static case, Mineur has shown that it is the same in the case where r and t are also time dependent.

b. Charged particle. If we consider a charged particle, we must take into account the electromagnetic field produced by the particle. Then instead of (5.1) we use

$$G_{\mu\nu} = \chi E_{\mu\nu}, \tag{5.5}$$

where $E_{\mu\nu}$ is the electromagnetic energy tensor endowed with a geometrical significance. If we postulate a maxwellian form of the electromagnetic field, we have ((5) p. 185).

$$E_{\mu\nu} = \tau_{\mu\nu} = \frac{1}{4} g_{\mu\nu} F_{\rho\sigma} F^{\rho\sigma} - F_{\mu\rho} F_{\nu\rho}, \tag{5.6}$$

with

$$F_{\mu\nu} = \partial_\mu \varphi_\nu - \partial_\nu \varphi_\mu. \tag{5.7}$$

In the electrostatic case, the only non-vanishing component of the field in polar coordinates is F_{14}. It follows then that

$$\tau_1^1 = -\tau_2^2 = -\tau_3^3 = \tau_4^4 = \frac{e^2}{r^4}$$

Substitution into (5.5) leads to the following form instead of (5.4)

$$\beta = r^2, \quad \gamma^{(1)} = \frac{1}{\alpha^{(1)}} = 1 - \frac{2m}{r} + \frac{4\pi e^2}{r^2}. \tag{5.8}$$

where χ has been set equal to 8π and $V = 1$.

If we do not limit ourselves to the static case and if we recall that the nonvanishing components of the field are F_{14} and F_{23} in polar coordinates, the solutions will not be identical to (5.8). In this case, the metric depends on time. Narlikar and Vaidya (69) have obtained the solution

$$\beta = r^2, \quad \alpha^{(1)} = \frac{1}{1 - \frac{2m}{r}}, \quad \sigma^{(1)} = \frac{\dot{m}^2}{m'^2\left(1 - \frac{2m}{r}\right)^2}\left(1 - \frac{2m}{r}\right), \tag{5.9}$$

where $\dot{m}$ and m' designate $\partial m/\partial t$ and $\partial m/\partial r$, and m is a function of r and of t.

The above solutions assume that the electric current vanishes, that is we can place ourselves outside the charge distribution and conserve the electromagnetic field.

2. DETERMINATION OF THE SPHERICALLY SYMMETRIC SOLUTION IN THE UNIFIED THEORY.

The field equations are

$$\text{II} \begin{cases} W_{\underline{\mu\nu}} = 0, \\ W_{\mu\nu\rho} = \partial_\mu \underset{\vee}{W}_{\nu\rho} + \partial_\rho \underset{\vee}{W}_{\mu\nu} + \partial_\nu \underset{\vee}{W}_{\rho\mu} = 0, \\ L_\rho = 0 \quad (\text{or } \partial_\rho \mathcal{G}^{\mu\rho} = 0) \end{cases}$$

$W_{\mu\nu}$ being the Ricci tensor formed with the connection $L^{\rho}_{\mu\nu}$ whose torsion $L_\rho = \underset{\vee}{L}^{\sigma}_{\rho\sigma}$ is zero. This connection is determined by the general solution (3.52, 3.56). It is sufficient to particularize this solution to the case where the $g_{\mu\nu}$ have the form which follows from spherical symmetry. Then (3.52 - 3.56) simplify considerably.

3. FORM OF THE SPHERICALLY SYMMETRIC SOLUTION.

a. Papapetrou's method.

The form of the $g_{\mu\nu}$ is the spherically symmetric case has been indicated by Papapetrou (71).

He assumes that the symmetric part $\gamma_{\mu\nu}$ of the affine connection has the same form as in general Relativity, namely

$$\gamma_{\mu\nu} = \begin{bmatrix} -\alpha & & & \\ & -\beta & & \\ & & -\beta \sin^2\theta & \\ & & & \sigma \end{bmatrix}, \tag{5. 10}$$

where α, β and σ are functions of r and eventually t.

The form of the antisymmetric part $\varphi_{\mu\nu}$ is determined from the following considerations. A 90° rotation about the z-axis will change the (x, y, z) coordinates of a point M into $x' = -y$, $y' = x$, $z' = z$, $t' = t$. Then

$$\varphi'_{12} = \varphi_{12}, \quad \varphi'_{23} = -\varphi_{31}, \quad \varphi'_{31} = \varphi_{23};$$

$$\varphi'_{14} = -\varphi_{24}, \quad \varphi'_{24} = \varphi_{14}, \quad \varphi'_{34} = \varphi_{34}.$$

It follows then that the components of $\varphi_{\mu\nu}$ for a point P on the z axis ($x'^{\mu} = x^{\mu}$) will reduce to φ_{12} and φ_{34} since $\varphi'_{\mu\nu} = \varphi_{\mu\nu}$. We can then go from $P_1(0, 0, z)$ to $P(x, y, z)$ by a rotation such that

$$x^2 + y^2 + z^2 = r^2.$$

In cartesian coordinates, the $\varphi_{\mu\nu}$ at the point P will be:

$$\varphi_{\mu\nu} = \begin{pmatrix} 0 & \frac{z}{r}v & -\frac{y}{r}v & \frac{x}{r}w \\ -\frac{z}{r}v & 0 & \frac{x}{r}v & \frac{y}{r}w \\ \frac{y}{r}v & -\frac{x}{r}v & 0 & \frac{z}{r}w \\ -\frac{x}{r}w & -\frac{y}{r}w & -\frac{z}{r}w & 0 \end{pmatrix} \tag{5. 11}$$

where v and w are any functions of r and t. Transforming to polar coordinates by

$$x = r \sin\theta \cos\varphi, \qquad y = r \sin\theta \sin\varphi, \qquad z = r\cos\theta, \tag{5.12}$$

we have

$$\varphi_{\mu\nu} = \begin{bmatrix} 0 & 0 & 0 & w \\ 0 & 0 & r^2 v \sin\theta & 0 \\ 0 & -r^2 v \sin\theta & 0 & 0 \\ -w & 0 & 0 & 0 \end{bmatrix} \tag{5.13}$$

The form of $g_{\mu\nu}$ obtained by Papapetrou will then be:

$$g_{\mu\nu} = \begin{bmatrix} -\alpha & 0 & 0 & w \\ 0 & -\beta & r^2 v \sin\theta & 0 \\ 0 & -r^2 v \sin\theta & -\beta \sin^2\theta & 0 \\ -w & 0 & 0 & \sigma \end{bmatrix} \tag{5.14}$$

Nearly all the work on spherically symmetric solutions admit this representation.

b. Vaidya's method.

The above derivation defines the $g_{\mu\nu}$ only in a radial direction. More generally, we could seek the form of $\varphi_{\mu\nu}$ at any point on a sphere which is undergoing an infinitesimal rotation about one of its diameters. This was developed by Vaidya (82), (83).

The components ξ^μ of an infinitesimal rotation which transform a sphere into itself are determined from Killing's equations and are:

$$\xi^1 = 0, \quad \xi^2 = A\cos(\varphi + B), \quad \xi^3 = -\frac{A \sin(\varphi+B)}{\mathrm{tg}\theta} + C,$$
$$\xi^4 = 0. \tag{5.15}$$

In an infinitesimal rotation, we have:

$$x'^\rho = x^\rho + \xi^\rho \,\delta\epsilon \tag{5.16}$$

The condition of spherical symmetry

$$g'_{\mu\nu}(x'^{\rho}) = g_{\mu\nu}(x'^{\rho}) \tag{5.17}$$

can then be written

$$\begin{aligned} g'_{\mu\nu}(x'^{\rho}) &= g_{\mu\nu}(x^{\rho}) + \frac{\partial g_{\mu\nu}}{\partial x^{\rho}} \xi^{\rho}\, \delta\epsilon \\ &= g_{\mu\nu}(x^{\rho}) - \left[g_{\mu\rho} \frac{\partial \xi^{\rho}}{\partial x^{\nu}} + g_{\rho\nu} \frac{\partial \xi^{\rho}}{\partial x^{\mu}} \right] \delta\epsilon . \end{aligned} \tag{5.18}$$

It follows then

$$g_{\mu\rho} \frac{\partial \xi^{\rho}}{\partial x^{\nu}} + g_{\rho\nu} \frac{\partial \xi^{\rho}}{\partial x^{\mu}} + \frac{\partial g_{\mu\nu}}{\partial x^{\rho}} \xi^{\rho} = 0 \tag{5.19}$$

where the ξ^{μ} are defined by (5. 15).

If we assume that the symmetric part of $g_{\mu\nu}$ has the form (5. 10)* then (5. 19) leads to

$$\varphi_{\mu\nu} = \begin{bmatrix} 0 & Pv & -Pu\sin\theta & Hh \\ -Pv & 0 & Ek\sin\theta & Qv \\ Pu\sin\theta & -Ek\sin\theta & 0 & -Qu\sin\theta \\ -Hh & -Qv & Qu\sin\theta & 0 \end{bmatrix}, \tag{5.20}$$

where **

$$w = A\sin(\varphi + B)\sin\theta + C\cos\theta, \tag{5.21}$$

$$\left\{ v = \frac{A\cos(\varphi + B)\, f(w)}{\sqrt{A^2 + C^2 - w^2}}, \quad u = \frac{A\sin(\varphi + B)\cos\theta - C\sin\theta}{\sqrt{A^2 + C^2 - w^2}} f(w), \right.$$

---

* Vaidya has proposed a polarized form of $\gamma_{\mu\nu}$ by assuming that they do no necessarily represent the metric tensor of space-time (cf. (82) p. 698).

** In this subsection, u, v, w, h and k are the same as the ones defined by Vaidya. They are not the same as the u, v, w, . . . of the rest of this work.

P, Q, E and H being arbitrary functions of r and t; h, k, and f being arbitrary functions of w.

The functions w, v, and u introduce a so-called polarization of the components, that is, the spherical symmetry condition can be satisfied either for all values of the parameters A, B and C which determine the rotation or for special values. The first alternative leads to a tensor field which is the solution proposed by Papapetrou. The second alternative leads to the polarized field since the $\varphi_{\mu\nu}$ depends on u, v and w which are functions of A, B and C. If we choose the axis of rotation (or axis of polarization) as the z-axis, we have

$$\varphi_{\mu\nu} = \begin{bmatrix} 0 & 0 & -P\sin\theta\, f(\theta) & H\, h(\theta) \\ 0 & 0 & E\sin\theta\, k(\theta) & 0 \\ +P\sin\theta\, f(\theta) & -E\sin\theta\, k(\theta) & 0 & -Q\sin\theta\, f(\theta) \\ -H\, h(\theta) & 0 & Q\sin\theta\, f(\theta) & 0 \end{bmatrix} \tag{5.22}$$

The form of these $\varphi_{\mu\nu}$ reduces to the Papapetrou form if P = Q = 0 and h = k = 1 (82).

If we carry out a transformation of coordinates (r, t) to (r', t') such that

$$P\left(\frac{\partial r}{\partial t'}\right) + Q\left(\frac{\partial t}{\partial t'}\right) = 0,$$

$$\sigma\left(\frac{\partial t}{\partial r'}\right)^2 + 2\alpha\left(\frac{\partial t}{\partial r'}\right)\left(\frac{\partial r}{\partial r'}\right) - \alpha\left(\frac{\partial r}{\partial r'}\right)^2 = 0,$$

we can define a system r't' said to be a Newtonian system in which

$$g_{\mu\nu} = \begin{bmatrix} 0 & 0 & P\sin\theta & \alpha + H \\ 0 & -\beta & E\sin\theta & 0 \\ -P\sin\theta & -E\sin\theta & -\beta\sin^2\theta & 0 \\ \alpha - H & 0 & 0 & \sigma \end{bmatrix} \tag{5.22'}$$

The equations

$$\partial_\rho\, \mathfrak{g}^{\mu\rho} = 0$$

imply the conditions

$$PEH = 0.$$

We can have then:

1. P = 0: these are the solutions discussed by Bonnor.
2. EH = 0: substitution of (5. 22') into $W_{\mu\nu} = \lambda\gamma_{\mu\nu}$ ($\lambda = 0$ or $\lambda \neq 0$) leads to P = 0. These are the solutions studied by Papapetrou and Wyman.

In short, the form (5. 22) proposed by Vaidya seems to reduce to one of the three forms studied by Papapetrou, Wyman and Bonnor (83). These are the cases 1, 2 and 3 discussed in this chapter.

4. AFFINE CONNECTION IN THE SPHERICALLY SYMMETRIC CASE.

Using polar coordinates

$$x^1 = r, \quad x^2 = \theta, \quad x^3 = \varphi, \quad x^4 = t, \tag{5. 3}$$

we assume that the tensor

$$g_{\mu\nu} = \gamma_{\mu\nu} + \varphi_{\mu\nu}$$

has the form (5. 14) given by Papapetrou

$$\gamma_{\mu\nu} = \begin{bmatrix} -\alpha & & & \\ & -\beta & & \\ & & -\beta\sin^2\theta & \\ & & & \sigma \end{bmatrix}, \quad \varphi_{\mu\nu} = \begin{bmatrix} 0 & 0 & 0 & w \\ 0 & 0 & u\sin\theta & 0 \\ 0 & -u\sin\theta & 0 & 0 \\ -w & 0 & 0 & 0 \end{bmatrix}, \tag{5. 23}$$

where

$$u = r^2 v.$$

The various determinants are

$$\gamma = -\alpha\sigma\beta^2 \sin^2 \theta, \tag{5.24}$$

$$\varphi = w^2 u^2 \sin^2 \theta, \tag{5.25}$$

$$g = -(\alpha\sigma - w^2)(\beta^2 + u^2) \sin^2 \theta. \tag{5.26}$$

Use of Eqs. (1.21) leads, when applied to this case, to

$$\left.\begin{aligned}
&\gamma^{\mu\nu} = \begin{bmatrix} \frac{-1}{\alpha} & & & \\ & \frac{-1}{\beta} & & \\ & & \frac{-1}{\beta \sin^2\theta} & \\ & & & \frac{1}{\sigma} \end{bmatrix}, \quad \varphi^{\mu\nu} = \begin{bmatrix} & & & \frac{1}{w} \\ & & \frac{1}{u\sin\theta} & \\ & \frac{-1}{u\sin\theta} & & \\ \frac{-1}{w} & & & \end{bmatrix}, \\
&\varphi^*_{\mu\nu} = \frac{\sqrt{-\gamma}}{2}\epsilon_{\mu\nu\rho\sigma}\gamma^{\rho\lambda}\gamma^{\sigma\tau}\varphi_{\lambda\tau} = \begin{bmatrix} 0 & 0 & 0 & \frac{u\sqrt{\alpha\sigma}}{\beta} \\ 0 & 0 & \frac{-\beta w}{\sqrt{\alpha\,\sigma}}\sin\theta & 0 \\ 0 & \frac{\beta w}{\sqrt{\alpha\sigma}}\sin\theta & 0 & 0 \\ \frac{-u\sqrt{\alpha\sigma}}{\beta} & 0 & 0 & 0 \end{bmatrix}
\end{aligned}\right\} \tag{5.27}$$

To determine the affine connection, we substitute the above values into Eqs. (3.52) -(3. 56) which express the affine connection in terms of the $g_{\mu\nu}$. The computations are carried out in Appendix IV. The results are summarized in the following table (57) (58).

Static condition	Non-static condition
Symmetric group $\Delta^{\rho}_{\underline{\mu\nu}}$.	
$\Delta^{1}_{11} = \frac{\alpha'}{2\alpha}$	$\Delta^{4}_{44} = \frac{\dot{\sigma}}{2\sigma}$
$\Delta^{1}_{22} = \frac{\Delta^{1}_{33}}{\sin^2\theta} = \frac{uB_1 - \beta A_1}{2\alpha}$	$\Delta^{4}_{22} = \frac{\Delta^{4}_{33}}{\sin^2\theta} = \frac{\beta A_4 - u B_4}{2\sigma}$
$\Delta^{1}_{44} = \frac{\alpha}{2\alpha}\, \partial_1 \mathrm{Log}\, \sigma \left(1 - \frac{w^2}{\alpha\sigma}\right)^2$	$\Delta^{4}_{11} = \frac{\alpha}{2\sigma}\, \partial_4 \mathrm{Log}\, \alpha \left(1 - \frac{w^2}{\alpha\sigma}\right)^2$
$\Delta^{4}_{\underline{14}} = \frac{1}{2}\, \partial_1 \mathrm{Log}\, \sigma \left(1 - \frac{w^2}{\alpha\sigma}\right)^2$	$\Delta^{1}_{\underline{14}} = \frac{1}{2} \partial_4 \mathrm{Log}\, \alpha \left(1 - \frac{w^2}{\alpha\sigma}\right)$
$\Gamma^{2}_{33} = -\sin\theta\cos\theta,$	$\Delta^{3}_{23} = \Delta^{3}_{32} = \frac{1}{\mathrm{tg}\,\theta}$
$\Delta^{2}_{\underline{12}} = \Delta^{3}_{\underline{31}} = \frac{A_1}{2}$ $\Delta^{2}_{\underline{34}} = -\Delta^{3}_{\underline{24}} \sin^2\theta = \frac{w B_1}{2\alpha} \sin\theta$	$\Delta^{2}_{\underline{24}} = \Delta^{3}_{\underline{34}} = \frac{A_4}{2}$ $\Delta^{2}_{\underline{24}} = -\Delta^{3}_{\underline{12}} \sin^2\theta = \frac{w B_4}{2\sigma} \sin\theta$
Antisymmetric group $\Delta^{\rho}_{\underset{\vee}{\mu\nu}}$.	
$\Delta^{1}_{23} = -\Delta^{1}_{32} = \frac{\beta B_1 + u A_1}{2\alpha} \sin\theta$	$\Delta^{4}_{23} = -\Delta^{4}_{32} = -\frac{\beta B_4 + u A_4}{2\sigma} \sin\theta$
$\Delta^{1}_{\underset{\vee}{14}} = \frac{\sigma}{2w} \partial_1 \mathrm{Log} \left(1 - \frac{w^2}{\alpha\sigma}\right)$	$\Delta^{4}_{\underset{\vee}{14}} = \frac{-\alpha}{2w}\, \partial_4 \mathrm{Log} \left(1 - \frac{w^2}{\alpha\sigma}\right)$
$\Delta^{2}_{\underset{\vee}{31}} = \Delta^{3}_{\underset{\vee}{12}} \sin^2\theta = -\frac{B_1}{2} \sin\theta$	$\Delta^{2}_{\underset{\vee}{34}} = -\Delta^{3}_{\underset{\vee}{24}} \sin^2\theta = -\frac{B_4}{2} \sin\theta$
$\Delta^{2}_{\underset{\vee}{24}} = \Delta^{3}_{\underset{\vee}{34}} = -\frac{w}{2\alpha} A_1$	$\Delta^{2}_{\underset{\vee}{12}} = -\Delta^{3}_{\underset{\vee}{31}} = \frac{w}{2\sigma} A_4$
	(5.28)

where

$$\left\{\begin{aligned} A_1 &= \frac{1}{2}\partial_1 \operatorname{Log}\beta^2\left(1+\frac{u^2}{\beta^2}\right) = \frac{uu' + \beta\beta'}{u^2+\beta^2}, \\ B_1 &= -\frac{\beta}{2u}\partial_1 \operatorname{Log}\left(1+\frac{u^2}{\beta^2}\right) = \frac{u\beta' - \beta u'}{u^2+\beta^2}; \end{aligned}\right. \tag{5.29}$$

$$\left\{\begin{aligned} A_4 &= \frac{1}{2}\partial_4 \operatorname{Log}\beta^2\left(1+\frac{u^2}{\beta^2}\right) = \frac{\dot{u}u + \dot{\beta}\beta}{u^2+\beta^2}, \\ B_4 &= -\frac{\beta}{2u}\partial_4 \operatorname{Log}\left(1+\frac{u^2}{\beta^2}\right) = \frac{u\dot{\beta} - \beta\dot{u}}{u^2+\beta^2}; \end{aligned}\right. \tag{5.30}$$

which is a generalization of Bonnor's notation to the non-static case. Here

$$\partial_1 = \frac{\partial}{\partial r}, \qquad \partial_4 = \frac{\partial}{\partial t}$$

and the primes indicate differentiation with respect to r, and the dot with respect to time.

The general condition for the existence of a solution (Chapter 3, § 6)

$$g(a^2 + b^2) \neq 0 \tag{3.57}$$

or

$$g\left[\left(2 - \frac{g}{\gamma} + \frac{6\varphi}{\gamma}\right)^2 - \frac{4\varphi}{\gamma}\left(3 - \frac{g}{\gamma} + \frac{\varphi}{\gamma}\right)^2\right] \neq 0 \tag{3.58}$$

becomes in our case

$$\left(1+\frac{u^2}{\beta^2}\right)\left(1-\frac{w^2}{\alpha\sigma}\right)\left[\left(1-\frac{u^2}{\beta^2}\right)^2 + \frac{4w^2u^2}{\alpha\sigma\beta^2}\right]\left[\left(1+\frac{w^2}{\alpha\sigma}\right)^2 + \frac{4w^2u^2}{\alpha\sigma\beta^2}\right] \neq 0. \tag{5.31}$$

We have deduced the affine connection and the conditions for existence of the general solution (3. 52) -(3. 56) and the conditions for existence (3. 58). However, if we limit ourselves to the purely static case, that is if we keep only the left hand side of (5. 28), we are led to the expressions computed by Bonner ((63) Eq. 2. 1) starting directly from $g_{\underset{+}{\mu}\underset{-}{\nu};\rho} = 0$. These equations simplify and can be resolved easily. Bonner also established directly the condition (5. 31) which we deduced from general existence conditions. The agreement with Bonnor's results is thus obtained in a special case and serves to show the validity of our solution.

5. THE DIFFERENTIAL EQUATIONS OF THE SPHERICALLY SYMMETRIC FIELD.

The affine connection Δ given by (5. 28) is now substituted into the expression

$$W_{\mu\nu} = \partial_\rho L^{\rho}_{\mu\nu} - \partial_\nu L^{\rho}_{\mu\rho} + L^{\lambda}_{\mu\nu} L^{\rho}_{\lambda\rho} - L^{\lambda}_{\mu\rho} L^{\rho}_{\lambda\nu} \qquad (5.\ 32)$$

$$\text{such as } L_\rho = L^{\sigma}_{\rho\sigma} = 0,$$

the connection L being the same as Δ with its torsion vector set equal to zero. Let $W^{(1)}_{\mu\nu}$, $W^{(4)}_{\mu\nu}$, and $W^{(14)}_{\mu\nu}$ be the parts of the components such that the superscripts designate:

(1) purely static
(4) purely dynamic
(14) mixed

The only non-vanishing components of $W_{\mu\nu}$ are (66):

	Static contribution $W^{1}_{\mu\nu}$.	Non-static contribution $W^{4}_{\mu\nu}$.
W_{11}	$- A'_1 + \frac{\alpha' A_1}{2\alpha} - \frac{A_1^2 + B_1^2}{2}$ $- \frac{1}{2}\, \partial_1^2 \,\mathrm{Log}\,\sigma \left(1 - \frac{w^2}{\alpha\sigma}\right)$ $+ \frac{1}{4}(\partial_1 \,\mathrm{Log}\,\sigma U)\left(\frac{\alpha'}{\alpha} - \partial_1 \,\mathrm{Log}\,\sigma U\right)$	$w^2\, \frac{A_4^2 + B_4^2}{2\sigma^2} + \dot\alpha \left(\frac{\sigma\dot\alpha - \alpha\dot\sigma + 2\alpha\sigma A_4}{4\alpha\sigma^2}\right)$ $+ (\partial_4 \,\mathrm{Log}\, U)\, \frac{\dot\alpha\sigma - 2\alpha\dot\sigma + 4\alpha\sigma A_4}{4\sigma^2}$ $+ \left(\frac{\alpha^2}{4w^2} - \frac{\alpha}{2\sigma}\right)(\partial_4 \,\mathrm{Log}\, U)^2$ $+ \frac{\alpha}{2\sigma}\, \partial_4^2 \,\mathrm{Log}\,\alpha U^2$
$W_{22} = \frac{W_{33}}{\sin^2\theta}$	$1 + B_1\, \frac{\beta B_1 + u A_1}{2\alpha} + \partial_1\left(\frac{u B_1 - \beta A_1}{2\alpha}\right)$ $+ \frac{u B_1 - \beta A_1}{4\alpha}\, \partial_1 \mathrm{Log}(\alpha\sigma - w^2)$	$\partial_4\left(\frac{\beta A_4 - u B_4}{2\sigma}\right) - B_4 \frac{\beta B_4 + u A_4}{2\sigma}$ $+ \frac{\beta A_4 - u B_4}{4\sigma}\, \partial_4 \,\mathrm{Log}\,(\alpha\sigma - w^2)$
W_{44}	$w^2 \frac{A_1^2 + B_1^2}{2\alpha^2} + \sigma' \left(\frac{\alpha\sigma' - \alpha'\sigma + 2\alpha\sigma A_1}{4\alpha^2\sigma}\right)$ $+ \partial_1 \mathrm{Log}\, U \left(\frac{\alpha\sigma' - 2\alpha'\sigma + 4\alpha\sigma A_1}{4\alpha^2}\right)$ $+ \left(\frac{\sigma^2}{4w^2} - \frac{\sigma}{2\alpha}\right)(\partial_1 \mathrm{Log}\, U)^2$ $+ \frac{\sigma}{2\alpha}\, \partial_1^2 \mathrm{Log}\ \sigma U^2$	$- A_4 + \frac{\dot\sigma A_4}{2\sigma} - \frac{A_4^2 + B_4^2}{2}$ $- \frac{1}{2}\, \partial_4^2 \mathrm{Log}\ \alpha U + \frac{\dot\sigma}{4\sigma} \partial_4 \,\mathrm{Log}\,\alpha U$ (5.33) $- \frac{1}{4}(\partial_4 \log \alpha U)^2$
W_{23} $\frac{v}{\sin\theta}$	$\partial_1\left(\frac{\beta B_1 + u A_1}{2\alpha}\right) - B_1 \frac{u B_1 - \beta A_1}{2\alpha}$ $+ \frac{\beta B_1 + u A_1}{4\alpha}\, \partial_1 \mathrm{Log}\ (\alpha\sigma - w^2)$ $w\, \frac{A_1^2 + B_1^2}{2\alpha}$	$- \partial_4 \left(\frac{\beta B_4 + u A_4}{2\sigma}\right) - B_4\, \frac{\beta A_4 - u B_4}{2\sigma}$ $- \frac{\beta B_4 + u A_4}{4\sigma}\, \partial_4 \mathrm{Log}(\alpha\sigma - w^2)$ $- w\, \frac{A_4^2 + B_4^2}{2\sigma}$
W_{14} v	$+ \frac{A_1\sigma w + w\sigma' - \sigma w'}{2w^2}\, \partial_1\ \mathrm{Log}\ U$ $+ \frac{\sigma}{2w}\, \partial_1^2\ \mathrm{Log}\ U$	$- \frac{A_4\alpha w + w\dot\alpha - \alpha\dot w}{2w^2}\, \partial_4\ \mathrm{Log}\ U$ $- \frac{\alpha}{2w}\, \partial_4^2\ \mathrm{Log}\ U$
$\underline{W_{14}}^{14}$	$-(A_1 A_4 + B_1 B_4) U + \frac{1}{2}\left(\frac{A_1\dot\alpha}{\alpha} + \frac{A_4\sigma'}{\sigma} - \dot A_1 - A'_4\right)$ $+ \frac{1}{2}\, \partial_1\partial_4 \mathrm{Log}\, U - \frac{1}{4}\left(3 - \frac{\alpha\sigma}{w^2}\right) \partial_1\ \mathrm{Log}\, U\, \partial_4\ \mathrm{Log}\, U$ $+ \frac{1}{2}\left(A_1 - \frac{\sigma'}{2\sigma}\right) \partial_4\ \mathrm{Log}\, U + \frac{1}{2}\left(A_4 - \frac{\dot\alpha}{2\alpha}\right) \partial_1\ \mathrm{Log}\, U$	

where

$$U = 1 - \frac{w^2}{2\sigma} \tag{5.34}$$

We must also impose on these equations the condition

$$\left\{\begin{array}{l} L_\rho = \underset{\vee}{L}{}^{\sigma}_{\rho\sigma} = 0, \\ \text{equivalent to} \\ \partial_\rho(\sqrt{-g}\, f^{\mu\nu}) = 0, \text{ that is to say } \partial_1(\sqrt{-g}f^{14}) = \partial_4(\sqrt{-g}f^{14}) = 0. \end{array}\right. \tag{5.35}$$

From (1.18), we have

$$f^{14} = \frac{\varphi}{g}\varphi^{14} + \frac{\gamma}{g}\gamma^{11}\gamma^{44}\varphi_{14} = \frac{-w}{\alpha\sigma - w^2}, \tag{5.36}$$

Thus (5.35) becomes:

$$\partial_1\left(\frac{\beta^2 + u^2}{\alpha\sigma - w^2}\right)^{\frac{1}{2}}(-w \sin\theta) = \partial_4\left(\frac{\beta^2 + u^2}{\alpha\sigma - w^2}\right)^{\frac{1}{2}}(-w \sin\theta) = 0. \tag{5.37}$$

The solution of this equation is:

$$\frac{\beta^2 + u^2}{\alpha\sigma - w^2}\, w^2 = k^2, \tag{5.38}$$

k being an arbitrary constant. Eq. (5.38) can also be written

$$(\beta^2 + u^2)\frac{1 - U}{U} = k^2 \quad \text{or} \quad U = \frac{\beta^2 + u^2}{\beta^2 + u^2 + k^2}. \tag{5.39}$$

The condition $\partial_\mu \mathfrak{g}^{\mu\rho} = 0$ is thus satisfied if

$$\frac{w^2}{\alpha\sigma} = \frac{k^2}{\beta^2 + u^2 + k^2}, \tag{5.40}$$

which is always satisfied by $k = 0$ if $w = 0$. If $w \neq 0$, (5.40) can be written as

$$1 - \frac{\alpha\sigma}{w^2} = c_1(\beta^2 + u^2) \quad \left(c_1 = -\frac{1}{k^2}\right).$$

In the case of the spherically symmetric solution where $W_{\mu\nu}$ has the form (5.33), the field equations (II) reduce to

$$\boxed{\begin{array}{lll} W_{11} = 0, & & \\ W_{22} = 0, & \left.\begin{array}{l} \partial_1 \underset{V}{W_{23}} = 0 \\ \partial_4 \underset{V}{W_{23}} = 0 \end{array}\right\} & \text{or } \underset{V}{W_{23}} = \text{const.} \\ W_{44} = 0, & & \\ W_{\underline{14}} = 0, & & \end{array}} \qquad (5.41)$$

with

$$\begin{array}{ll} & \text{either } w = 0, \\ \text{or} & \dfrac{w^2}{\alpha\sigma} = \dfrac{k^2}{k^2 + \beta^2 + u^2}. \end{array} \qquad (5.40)$$

In addition the two equations $\partial_2 \underset{V}{W_{14}} = 0$ and $\partial_3 \underset{V}{W_{14}} = 0$ which occur in (II) are automatically satisfied since $\underset{V}{W_{14}}$ is only a function of r and t.

B. THE VARIOUS FORMS OF THE SPHERICALLY SYMMETRIC SOLUTIONS.

6. THE VARIOUS CASES OF RESOLUTION OF THE DIFFERENTIAL EQUATIONS LEADING TO A SPHERICALLY SYMMETRIC SOLUTION.

The equation

$$\partial_\rho \mathfrak{g}^{\mu\rho} = 0$$

has given the following conditions:

$$\text{either} \qquad w = 0; \qquad (5.40a)$$

$$\text{or} \qquad \frac{w^2}{\alpha\sigma} = \frac{k^2}{k^2 + \beta^2 + u^2}. \qquad (5.40b)$$

Three cases are distinguishable:

Case 1: $(w \neq 0,\ u = 0)$ this leads to

$$\frac{w^2}{\alpha\sigma} = \frac{k^2}{k^2 + \beta^2}.$$

Case 2: $(u \neq 0,\ w = 0)$ (5. 40) is automatically satisfied.

Case 3: $(w \neq 0,\ u \neq 0)$ the condition is:

$$\frac{w^2}{\alpha\sigma} = \frac{k^2}{\beta^2 + u^2 + k^2}$$

We shall treat these three cases separately.

Case 1: The particular case $u = 0,\ w \neq 0$.

7. PAPAPETROU'S SOLUTION (71).

In the particular case where $u = 0$, we have

$$\underset{V}{W_{23}} \equiv 0 \qquad (5.\ 42)$$

The field equations (5. 41) are:

$$\begin{cases} W_{11} = 0 \\ W_{22} = 0 \\ W_{44} = 0 \\ W_{\underline{14}} = 0 \end{cases} \qquad \frac{w^2}{\alpha\sigma} = \frac{k^2}{k^2 + \beta^2}. \qquad (5.\ 43)$$

The equation $W_{\underline{14}} = 0$ enters only in the non-static case. On the other hand, from the identity relations, one of the functions α, β, σ or w is arbitrary. Let

$$\beta = r^2.$$

We then have

$$A_1 = \frac{\beta'}{\beta}, \qquad B_1 = A_4 = B_4 = 0, \quad U = \frac{\beta^2}{k^2 + \beta^2} \qquad (5.\ 44)$$

and $W_{\underline{14}} = 0$ is always satisfied (cf. (5.53) if

$$\alpha = 0$$

of if

$$A_1 - \frac{1}{2}\,\partial_1 \text{ Log } U = 0.$$

From the value of U in (5. 44), it follows that $\dot{\alpha} = 0$ and the field equations must be identical in the static and the non-static cases (cf. Mavrides (66)).

From (5. 33), we have

$$\left.\begin{aligned}
W_{11} &= -\frac{1}{2}\partial_1^2 \text{ Log } \sigma - \frac{1}{4}\partial_1 \text{ Log}\,\sigma\,\partial_1 \text{ Log}\frac{\sigma}{\alpha} + \frac{1}{r}\partial_1 \text{ Log } \alpha \\
&\qquad - \frac{1}{2}\partial_1^2 \text{ Log}\,U + \frac{1}{4}\partial_1 \text{ Log}\,U\,\partial_1 \text{ Log}\frac{\alpha}{2}\,\sigma\,U = 0, \\
W_{22} &= 1 + \frac{r}{\alpha}\left(\frac{1}{2}\partial_1 \text{ Log}\frac{\alpha}{\sigma} - \frac{1}{r}\right) - \frac{r}{2\alpha}\,\partial_1 \text{ Log } U = 0, \\
\frac{\alpha}{\sigma}\,W_{44} &= \frac{1}{2}\partial_1^2 \text{ Log } \sigma + \frac{1}{4}\partial_1 \text{ Log } \sigma\,\partial_1 \text{ Log }\frac{\sigma}{\alpha} + \frac{1}{r}\partial_1 \text{ Log } \sigma \\
&\qquad + \partial_1^2 \text{ Log } U + \frac{1}{4}\partial_1 \text{ Log}\,U\,\partial_1 \text{ Log}\frac{\sigma^2 r^8}{\alpha U w^2} \\
&\qquad + \frac{2w^2}{\alpha\sigma r^2} = 0,
\end{aligned}\right\} \tag{5.45}$$

with

$$U = \frac{\beta^2}{k^2 + \beta^2}\,. \tag{5. 46}$$

We now compute $W_{11} + \frac{\alpha}{\sigma}\,W_{44}$. This yields:

$$\alpha\sigma = 1 + \frac{k^2}{\beta^2} = 1 + \frac{k^2}{r^4}\,. \tag{5. 47}$$

We can then deduce from (5. 44)

$$\left\{\begin{aligned}
U &= 1 - \frac{w^2}{\alpha\sigma} = \frac{\beta^2}{k^2 + \beta^2}, \\
w^2 &= \alpha\sigma\,\frac{k^2}{k^2 + \beta^2} = \frac{k^2}{\beta^2}.
\end{aligned}\right. \tag{5. 48}$$

On the other hand, the equations $W_{11} = 0$ and $W_{22} = 0$ reduce to a single equation since $\partial_1 W_{22} \equiv \frac{2r}{\alpha} W_{11}$. We thus have a single equation to determine α or σ. If we use $W_{22} = 0$, we have

$$1 + \frac{r}{2\alpha} \partial_1 \operatorname{Log} \alpha\sigma - \frac{r}{\alpha} \partial_1 \operatorname{Log} \sigma - \frac{1}{\alpha} - \frac{4}{2\alpha} \partial_1 \operatorname{Log} U = 0. \tag{5.49}$$

Taking into account the values of $\alpha\sigma$ and U as given by (5. 47), (5. 34) and (5. 48) and letting

$$e^{2n} = \frac{\sigma}{1 + \frac{k^2}{\beta^2}}$$

(5. 49) can be written in the form

$$2n' e^{2n} r + e^{2n} = e^{2n} (2rn' + 1) = V^2 \tag{5.50}$$

whose solution is:

$$e^{2n} = V^2 - \frac{2m}{r}. \tag{5.51}$$

In the case of $u = 0$, the spherically symmetric solution is

$$w = \frac{k}{r^2},$$

$$\beta = r^2, \qquad \alpha = \frac{1}{1 - \frac{2m}{V^2 r}}, \qquad \sigma = \left(V^2 - \frac{2m}{r}\right)\left(1 + \frac{k^2}{r^4}\right). \tag{5.52}$$

This was the solution determined by Papapetrou (cf. (71) pp. 74-75, Eq. (15) and (16)). If $r \to \infty$, we have

$$\alpha \to 1, \quad \beta \to r^2, \quad \sigma \to V^2, \quad w \to 0 \tag{5.53}$$

which shows that (5. 52) satisfies the condition that, as $r \to \infty$, the affine connection is euclidean.

8. CRITIQUE OF THE SOLUTION.

The first criticism of this solution lies in the fact that, as r becomes large, (5.52) does not agree with (5.8) which results from the introduction of Maxwell's tensor $\tau_{\mu\nu}$ into the equations of General Relativity. This may be inconvenient in that one would expect that, at large distances from the sources, the results of the unified theory should reduce to the results of the general theory of relativity combined with Maxwell's theory, even though the results are different near the sources. We think that this conclusion is wrong since we believe, as suggests Papapetrou, that the separation from Maxwell's theory must take place earlier.

Following Born let us imagine for a moment that, in a purely euclidean electromagnetic theory, we can substitute for the description of field due to point charges, the definition of a "free current density" with the aid of an electric field which remains finite as $r \to 0$. As in Born's theory, this density ρ will not be a function of the coordinates but will be expressed in terms of the fields. In this case, the asymptotic condition $\rho = 0$ can be realized only if the fields vanish. In a geometrical interpretation, this is equivalent to saying that the space is euclidean. It follows then that to a weak field would correspond a weak but non-vanishing value of ρ and could not possibly go over into the maxwellian description which gave (5.8), since this description is equivalent to placing ourselves outside the charges to cancel the current. In other words, the non-euclidean character of a point in space will be tied, not to the presence of the charges in the neighborhood of the point, but to the existence, at this point, of a density ρ which is a function of the field. The results obtained from such a theory could not go over into other theories which, in the limit, are valid only outside the sources. This limit corresponds here to a euclidean space.

We shall try to crystallize these considerations by examining the correspondence between Einstein's and Born's theories.

The second objection to the solution (5. 52) is its dependence on two arbitrary constants m and k. This leads one to think that the identification of the metric with (5. 52) is arbitrary. In fact, Wyman (85) has shown that it is possible to choose a metric which with the help of (5. 52), —which depends on two constants—goes over into Schwarzschild's solution—which depends on one constant.

Adopting Wyman's definition of the metric,

$$a_{\mu\nu} = \gamma_{\mu\nu} + q_\mu q_\nu, \tag{5. 54}$$

where

$$q_\mu = \frac{g_{\mu\rho} h^{\rho\sigma} u_\sigma}{\sqrt{1 + \frac{1}{2}\varphi_\rho f^{\rho\sigma}}} \tag{5. 55}$$

with

$$u_\mu = \frac{h_\mu}{\sqrt{g^{\rho\sigma} h_\rho h_\sigma}} = \frac{h_\mu}{\sqrt{h^{\rho\sigma} h_\rho h_\sigma}}, \quad h_\mu = \gamma_{\lambda\tau} f^{\sigma\lambda} \underset{V}{\Gamma^\tau_{\sigma\mu}}, \tag{5. 56}$$

we obtain, with the help of the values (5. 52) for $g_{\mu\nu}$, the following results:

The only nonvanishing components of the affine connection corresponding to (5. 52) are:

$$\underset{V}{\Delta^2_{24}} = \underset{V}{\Delta^3_{34}} = \frac{w}{rg_{11}}, \qquad \underset{V}{\Delta^1_{14}} = \frac{-2w}{rg_{11}} \tag{5. 57}$$

This leads to the following values of h_μ, u_μ and q_μ as defined in (5. 55) and (5. 56)

$$h_\mu = \left[\frac{-k^2}{r^5},\ 0,\ 0,\ 0\right], \tag{5. 58}$$

$$u_\mu = \left[\frac{-1}{\sqrt{g^{11}}},\ 0,\ 0,\ 0\right], \quad q_\mu = \left[0,\ 0,\ 0,\ \frac{g_{41}\sqrt{g^{11}}}{(1+g_{14}^2)^{\frac{1}{2}}}\right] \tag{5. 59}$$

We can then deduce from (5. 54) the following values for $a_{\mu\nu}$:

$$a_{11} = \frac{-1}{1 - \frac{2m}{r}}, \qquad a_{22} = -r^2, \quad a_{33} = -r^2 \sin^2\theta,$$

$$a_{44} = 1 - \frac{2m}{r}. \tag{5. 60}$$

Thus, we can define a particular metric $a_{\mu\nu}$ which may reduce to the Schwarzschild solution, which depends on one arbitrary constant; k, which is tied to the existence of the electromagnetic field will not enter in the determination of the metric.

Despite the artificial character of (5. 54), Wyman's remark is quite important in that it shows the arbitrary character in the definition of the "true" metric and the electromagnetic field. Their identification with the symmetric and antisymmetric parts of $g_{\mu\nu}$ is not automatic. A different choice (such as (5. 54), neither more nor less arbitrary, could lead to results unacceptable in a unified theory.

At this stage of the theory, we will prudently conclude that the fields $g_{\mu\nu}$ are related to the metric and the electromagnetic field but that the nature of this relation is still open. We shall discuss this further in Chapter 6.

Case 2:

9. WYMAN'S SOLUTION:

We limit ourselves to the static case. The differential equations (5. 41) reduce to:

$$\begin{cases} W_{11} = 0, \\ W_{22} = 0, \\ W_{44} = 0, \\ \partial_1 \underset{\vee}{W_{23}} = 0, \qquad \text{or} \quad \underset{\vee}{W_{23}} = \text{const.} \end{cases} \tag{5. 61}$$

The values of W^{11}, W_{22}, W_{44} and $\underset{\vee}{W_{23}}$ are given by the left hand side of (5. 33) in which we set $w = 0$. Writing A, and B, instead of we have

$$A' + \frac{1}{2}(A^2 + B^2) - \frac{1}{2} A \frac{\alpha'}{\alpha}$$
$$+ \frac{1}{2}\left[\partial_1^2 \operatorname{Log} \sigma - \frac{\alpha'}{\alpha} \partial_1 \operatorname{Log} \sigma - \frac{1}{2}(\partial_1 \operatorname{Log} \sigma)^2\right] = 0. \tag{5. 62}$$

$$1 + \frac{B}{2\alpha}(\beta B + uA) + \partial_1\left(\frac{uB - \beta A}{2\alpha}\right) + \frac{uB - \beta A}{4\alpha}\partial_1 \operatorname{Log}\alpha\sigma = 0, \tag{5.63}$$

$$\frac{1}{2}\partial_1 \operatorname{Log}\sigma\left(\frac{\sigma'}{\sigma} - \frac{\alpha'}{\alpha} + 2A\right) + \partial_1^2 \operatorname{Log}\sigma = 0, \tag{5.64}$$

$$\partial_1\left(\frac{\beta B + uA}{2\alpha}\right) - B\,\frac{uB - \beta A}{2\alpha} + \frac{\beta B + uA}{4\alpha}\partial_1 \operatorname{Log}\alpha\sigma = 0. \tag{5.65}$$

The solution of these equations has been carried out by Wyman (85). In particular, we have

$$\sigma' = \frac{2m\sqrt{\alpha\sigma}}{\sqrt{u^2 + \beta^2}} \tag{5.66}$$

Introducing $e^q = u + i\beta$, we have two types of solution depending on whether or not m vanishes. In Case 1, we had set $\beta = r^2$. Since one of the functions, α, β, or σ is arbitrary, we shall choose σ to be so and express α, β and u in terms of it. We then have (cf (85) p. 430, Eq. 120):

If $m \neq 0$

$$u + i\beta = \frac{4m^2 h}{\left[e^a \gamma^{h/2} + e^{-a}\gamma^{-h/2}\right]\sigma(c + i)}, \tag{5.67}$$

$$\alpha = \frac{(\sigma')^2(u^2 + \beta^2)}{4m^2\sigma},$$

with

$$h = 1 + ih_1.$$

In these expressions, σ is an arbitrary function of r. m, c and h are real constants while a is complex. A particular solution deduced from (5.67) for the choice

$$h_1 = 0, \quad e^{2a} = -1, \quad \sigma = 1 - \frac{2m}{r}.$$

is

$$\beta = r^2, \quad u = -cr^2.$$

This particular solution was obtained by Papapetrou for $w = 0,\ u \neq 0$ before Wyman obtained the solution (5.67).

If $m = 0$ (cf (85) p. 432, Eq. (2.7) and (2.8)), we have two sets of solutions:

Either

$$\left.\begin{aligned} u + i\beta &= \frac{h \operatorname{sech}^2(\sqrt{h}x + a)}{c + i}, \\ \sigma = 1, \quad \alpha &= (u^2 + \beta^2)\left(\frac{dx}{dr}\right)^2; \end{aligned}\right\} \tag{5.68}$$

or

$$\left.\begin{aligned} u + i\beta &= \frac{i - c}{(c^2 + 1)(x + a)^2}, \\ \sigma = 1, \quad \alpha &= (u^2 + \beta^2)\left(\frac{dx}{dr}\right)^2; \end{aligned}\right\} \tag{5.69}$$

where h and a are complex constants and $x = x(r)$.

10. THE LIMITING CONDITIONS.

It is interesting to note that the limit in conditions satisfied by (5.67) differ as to when the solution is expressed in cartesian or polar coordinates (cf.(85) p. 000). We must have:

$$\left.\begin{aligned} &\alpha \to 1, \qquad \sigma \to 1, \quad \beta \to \infty \\ &\varphi_{\mu\nu} \to 0 \end{aligned}\right\} \text{when } r \to \infty. \tag{5.70}$$

a. Weak conditions. In a cartesian system where the $\varphi_{\mu\nu}$ have the form (5.11), we have

$$\left.\begin{aligned} v &\to 0 \\ w &\to 0 \end{aligned}\right\} \text{when } r \to \infty. \tag{5.71}$$

b. Strong conditions. In polar coordinates, we have

$$\left.\begin{aligned} u = r^2 v &\to 0 \\ w &\to 0 \end{aligned}\right\} \text{when } r \to \infty. \tag{5.72}$$

The strong conditions applied to the solution in $m \neq 0$* lead to a solution with one arbitrary constant. In fact it reduces to the Schwarzschild solution.

The weak conditions applied to the same solution lead to

$$u + i\beta = 4m^2\ ih \frac{\sigma^{\sqrt{h}} - 1}{\left(\sigma^{\sqrt{h}} - 1\right)^2} \qquad (h = 1 + ih_1) \quad (5.73)$$

which depends on two arbitrary real constants m and h_1.

Wyman uses the strong conditions which unite the two particular cases ($u = 0$, $w \neq 0$, studied by Papapetrou and $w = 0$, $u \neq 0$, studied by Wyman). The solutions can always degenerate into a Schwarzschild solution depending on one arbitrary constant by either using the case $w \neq 0$, $u = 0$ with a choice of the appropriate coordinate system or in the case $u \neq 0$, $w = 0$ with use of the strong conditions at the limits as $u \to 0$.

The above argument is not very convincing since the particular cases $u = 0$ and $w = 0$ do not enter at all in the search for spherically symmetric solution and are not introduced in the same way (see paragraph 5). Furthermore, the reduction to the Schwarzschild solution is carried out in a completely different way.

We shall now examine the general case $u \neq 0$, $w \neq 0$ with both φ_{14} and φ_{23} present.

Case 3:

11. EQUATIONS FOR THE WEAK SYSTEM AND THE STRONG SYSTEM.

If we restrict ourselves to the static case, the systems (II) reduce to (5.40, 5.41)

$$\left.\begin{aligned} W_{11} &= 0, \\ W_{22} &= 0, \qquad \underset{\vee}{W_{23}} = \text{const.}, \\ W_{44} &= 0; \end{aligned}\right\} \quad (5.74)$$

---

* We note that the application of the strong conditions to the solutions for $m = 0$ will lead to a disappearance of all arbitrary constants (Wyman (85) p. 439).

with the condition (5. 40b)

$$\frac{w^2}{\alpha\sigma} = \frac{k^2}{k^2 + \beta^2 + u^2}. \qquad (5.40b)$$

We substitute here the $W_{\mu\nu}$ by their values from the left-hand side of (5. 33) and solve (5. 74) as Wyman did for a particular case. To our knowledge, this has not as yet been carried out.

Bonnor (63) has resolved Einstein's equations for the strong system (cf.(F) p. 00)

$$\begin{aligned} W_{\mu\nu} &= 0, \\ \underset{\vee}{W_{\mu\nu}} &= 0, \end{aligned} \qquad (5.75)$$

with the condition

$$L_\rho = \underset{\vee}{L^\sigma_{\rho\sigma}} = 0 \text{ equivalent to } \partial_\mu \mathfrak{g}^{\rho\mu} = 0$$

which is (5. 40) for the static case with spherical symmetry.

Then two possibilities differ by the equations (cf. p. 75).

$$\underset{\vee}{W_{23}} = \text{const.} \quad \partial_2 W_{14} = \partial_3 W_{14} \equiv 0 \text{ for system } 1^\circ;$$

$$\underset{\vee}{W_{23}} = 0, \quad \underset{\vee}{W_{14}} = 0 \text{ for system } 2.$$

We have not examined the solutions of the system (5. 75) until now: the reason is given by the deduction of the equations of the theory in Chapter 2 (cf. p. 28-31). We shall anyway summarize in the general Case 3, the solution of (5. 75) obtained by Bonnor (63). The method he used may be analogous to the resolution of the weak system (II) in the general case.

12. THE SOLUTION OF THE STRONG SYSTEM IN THE CASE $u \neq 0$, $w \neq 0$.

The equations that Bonnor attempts to solve are

$$\left.\begin{aligned} W_{11} &= 0, \\ W_{22} &= 0, \\ W_{44} &= 0, \\ \underset{\vee}{W_{23}} &= 0, \\ \underset{\vee}{W_{14}} &= 0, \end{aligned}\right\} \qquad (5.75)$$

with the condition

$$\underset{\vee}{L^{\rho}_{4\rho}} = 0.$$

This condition can always be expressed by (5. 40b)

$$\frac{w^2}{\alpha\sigma} = \frac{k^2}{k^2 + \beta^2 + u^2} \qquad (5.40)$$

which Bonnor writes

$$1 - \frac{\alpha\sigma}{w^2} = c_1(\beta^2 + u^2) \qquad \left(c_1 = -\frac{1}{k^2}\right) .$$

The results he obtains are:

a. Real: two possible solutions (cf.(63) p. 430-431, Eqs. (2.26) - (2.27))

(1)	(2)
$u + i\beta = -ih\,\mathrm{sech}^2(\sqrt{h}\,x + a)$	$u + i\beta = i(x + a)^{-2}$
$\alpha = (u^2 + \beta^2)\left(\frac{dx}{dr}\right)^2$	$\alpha = (u^2 + \beta^2)\left(\frac{dx}{dr}\right)^2$

$$\sigma = c + \frac{\mathcal{L}^2}{u^2 + \beta^2}$$

$$w = \pm \mathcal{L}\frac{dx}{dr}$$

(5.

where $h = i h_0$, h_0 and c are real constants; $\mathcal{L}$ can be real or imaginary, a is a complex constant and x is an arbitrary function of r.

b. Imaginary: two possible solutions (loc. cit. p. 431, Eqs. (2.28) - (2.29).

(1)

$$u = -\frac{ih}{2}[\ \mathrm{sech}^2(\sqrt{hx} + a) + \mathrm{sech}^2(\sqrt{-h}\,x + b)]$$

$$\beta = \frac{h}{2}[\ \mathrm{sech}^2(\sqrt{-h}\,x + b) - \mathrm{sech}^2(\sqrt{h}\,x + a)^2$$

$$\alpha = (u^2 + \beta^2)\ \frac{dx}{dr}^{\ 2}$$

(2)

$$u = \frac{i}{2}[(x + a)^{-2} - (x + b)^{-2}]$$

$$\beta = \frac{1}{2}[(x + a)^{-2} + (x + b)^{-2}]$$

$$\alpha = (u^2 + \beta^2)\left(\frac{dx}{dr}\right)^2$$

$$\sigma = c + \frac{\mathcal{L}^2}{u^2 + \beta^2}$$

$$w = \pm\mathcal{L}\frac{dx}{dr} \qquad (5.77)$$

where h and c are real constants, a and $\mathcal{L}$ are complex constants and x an arbitrary function of r.

13. THE CONDITIONS AT THE LIMITS.

We adopt the following limits.

$$\alpha \to 1, \quad \beta \to r^2, \quad \sigma \to 1, \quad v = \frac{u}{r^2} \to 0, \quad w \to 0,$$

when $r \to \infty$.

An examination of these limits shows that

$$x = \frac{1}{r}$$

will satisfy them. It follows then that the sets a and b reduce to (loc. cit., pp. 431, 432, Eqs. (3.2), (3.3) and (3.4))

a(1)	b(1)	a(2) and b(2)
$f = \frac{2h_0(1 - \cosh sx \cos sx)}{(\cosh sx - \cos sx)^2}$	$f = \frac{ih}{2}\left(\frac{1}{\cosh^2 s'x} - \frac{1}{\cos^2 s'x}\right)$	$f = 0$
$\beta = \frac{s^2 \sinh sx \sin sx}{(\cosh sx - \cos sx)^2}$	$\beta = \frac{\lvert h \rvert}{2}\left(\frac{1}{\cosh^2 s'x} + \frac{1}{\cos^2 s'x}\right)$	$\beta = \frac{1}{x^2}$
$\alpha = s^4 x^4 (\cosh sx - \cos sx)^{-2}$	$\alpha = h^2 x^4 \frac{1}{\cosh^2 s'x \cos^2 s'x}$	$\alpha = 1$ (5.78)
$\sigma = 1 + \mathcal{L}^2 s^{-4}(\cosh sx - \cos sx)^2$	$\sigma = 1 + \frac{l^2}{h^2} \sinh^2 s'x \sin^2 s'x$	$\sigma = 1 + l^2 x^4$
$w = \pm \mathcal{L} x^2$	$w = \pm l x^2$	$w = \pm l x^2$

where

$$s = \sqrt{2h_0} \quad \text{if u is real and h imaginary}$$

$$s' = \sqrt{|h|} \quad \text{if u is imaginary and h real.}$$

An examination of the solutions a (1) and b (1) of (5. 78) shows that if u does not vanish, then u will approach a finite limit as $r \to \infty$. The strong condition given by Wyman cannot be realized except in Case 2. It does not lead to a valid construction of a field $u \neq 0$. It is improbable but not certain that equations (5. 74) for the weak system will not lead to a more encouraging result in Case 3. Furthermore, (5. 78) cannot represent the properties of a change and mass distribution. In particular, no arbitrary constant in (5. 78) can correspond to the mass of a particle. It would therefore seem that the case $u \neq 0$, $w \neq 0$ does not lead to a satisfactory solution for the strong system. It may be that more satisfactory conclusions would result if a similar resolution is applied to the weak system as expressed by (5. 74).

14. CHOICE OF A SPHERICALLY SYMMETRIC SOLUTION.

The possibilities due to the choice of a spherically symmetric solution are the following:

I. One could, in the particular case $u = 0$, $w \neq 0$, adopt Papapetrou's solution (5.52). In this case one should note that the solution as given by (5.52) assumes that $u = 0$ and conserves only the field component φ_{14}. This, coupled, with the definition of the current as given in Chapter 4 (Section 2) would tie φ_{14} to a magnetic field and (5.52) would then correspond to the existence of isolated magnetic poles which is unacceptable. We shall discuss this objection in Chapter 6 (Section 1). We think that it can be removed if we define the current in such a way as to the φ_{14} to the electric field and to represent the charges by the solution $u = 0$, $w \neq 0$.

In addition, a certain arbitrariness exists in the choice of the metric and the electromagnetic field. Here, it seems again possible to remove some of this arbitrariness by considerations which will be discussed in Chapter 6 (Section 1). The removal of these two objections would then render the solution (5.52) a perfectly good one for the static case.

II. One could try to solve the equations for the weak system in the general case $u \neq 0$, $w \neq 0$. One obtains then a spherically symmetric solution tied to both types of fields.

III. All solutions discussed in this chapter assume:

a. the choice of a purely static solution

b. $\mathfrak{g}^{\mu}$ or $\underset{\vee}{L}{}^{\sigma}_{\rho\sigma} = 0$ which results automatically from a variational principle applied to the density $\mathfrak{H} = \mathfrak{G}^{\mu\nu} R_{\mu\nu}$ where $R_{\mu\nu}$ is the Ricci tensor.

c. The form (5.14) for the $g_{\mu\nu}$

These assumptions could be enlarged by

a. Consideration of the non-static case (this will go into the static case for $u = 0$). This could eventually be tied to the assumption of magnetic poles due to the creation of a magnetic field by rotating matter.

b. Deduction of the equations of the theory from a density $\mathfrak{H}^{(a)}$ formed with a tensor $R_{\mu\nu}^{(a)}$ other than the Ricci tensor. In particular if we use $R^{(1)}$, $R^{(2)}$, or $R^{(3)}$, the condition

$$\mathfrak{g}^{\mu} = 0, \qquad L_{\mu} = 0$$

does not follow from the variational principle and can be ignored. The spherically symmetric solutions would then involve less stringent assumptions.

c. Assuming a different form for the $g_{\mu\nu}$ for example to the conditions (5.19) suggested by Vaidya (83) (it seems that this should be associated with $\mathfrak{g}^{\mu} \neq 0$ in view of Vaidya's recent conclusions).

6

The Field and the Sources

A. CHOICE OF THE METRIC AND FIELDS IN THE UNIFIED THEORY.

The separation of the fundamental tensor into symmetric and antisymmetric components

$$g_{\mu\nu} = \gamma_{\mu\nu} + \varphi_{\mu\nu},$$

$$g^{\mu\nu} = h^{\mu\nu} + f^{\mu\nu}$$

leads to an indeterminacy in the choice of the metric and the fields. We could—at least in particular cases—compare the results of the solution of $\gamma_{\mu\nu}$ and $h^{\mu\nu}$ in unified theory to those of general Relativity. However, to do this, we must know which fields or combination of fields are susceptible of representing the "true" metrics. We could postulate, a priori, that the "true" metric $a_{\mu\nu}$ must be a function of $g_{\mu\nu}$, $\Gamma^{\rho}_{\mu\nu}$ that is of $g_{\mu\nu}$ and its derivatives. Keeping in mind the pertinent remarks of Wyman (85), we could assume that the metric has the form

$$a_{\mu\nu} = l_{\mu\nu}(g_{\rho\sigma}, \Gamma^{\lambda}_{\rho\sigma}) = l_{\mu\nu}(g_{\rho\sigma}, \partial_{\lambda}g_{\rho\sigma}) \qquad (6.1)$$

provided that $l_{\mu\nu}$ is such that $a_{\mu\nu}$ still represents a symmetric tensor. In particular, Wyman has shown that a certain choice of the metric, although quite artificial, will permit us to write Papapetrou's solution (static spherically symmetric solution with $u = 0$, $w \neq 0$) in the form of a Schwarzschild solution with one arbitrary constant. This example illustrates the fact that the choice of a metric in Einstein's unified theory is an important but unresolved question.

A desire for simplicity has led most authors to $i\gamma_{\mu\nu}$ ($g_{\mu\nu}$ in Einstein's notation) to the metric. However as proposed by Lichnerowicz (cf. (10), p. 1288), it is as simple and more justifiable to associate $h^{\mu\nu}$ with the metric ($h^{\mu\nu} = g^{\underline{\mu\nu}}$). In this fashion the waves φ satisfy the characteristic equation:

$$\Delta_1\varphi = g^{\mu\nu}\,\partial_\mu\varphi\partial_\nu\,\varphi = h^{\mu\nu}\partial_\mu\,\varphi\,\partial_\nu\varphi = 0. \qquad (6.2)$$

In any case the metric can be defined by the $\gamma_{\mu\nu}$ or $h^{\mu\nu}$ up to an invariant.

To guide ourselves in the choice of the metric and the fields, we can try to adopt certain definitions which will permit us to find in the unified theory certain results of the Born-Infeld nonlinear electromagnetic theory. We shall give a short summary of the theory before discussing the above question.

B. PRINCIPLES OF THE BORN-INFELD THEORY (122).

We define in a euclidean space, a coordinate system by introducing the metric $a_{\mu\nu}$. We shall adopt the convention that a bar under the letters as to whether the index has been raised or lowered by use of $a_{\mu\nu}$. If $s_{\mu\nu}$ and $s^{\underline{\mu\nu}} = a^{\mu\rho}a^{\nu\sigma}s_{\rho\sigma}$ represent the covariant and contravariant components of the electromagnetic field, then the two invariants of Maxwell's theory are

$$F = \frac{1}{2}\,s_{\mu\nu}\,s^{\underline{\mu\nu}} = \frac{1}{2}\,a^{\mu\rho}\,a^{\nu\sigma}\,s_{\mu\nu}\,s_{\rho\sigma}, \qquad (6.3)$$

$$G = \frac{1}{4}\,s_{\mu\nu}s^{*\mu\nu} = \frac{1}{8\sqrt{-a}}\,\epsilon^{\mu\nu\rho\sigma}\,s_{\mu\nu}\,s_{\rho\sigma} = \frac{s}{\sqrt{-a}}, \qquad (6.4)$$

where s and a are the determinants of $s_{\mu\nu}$ and $a_{\mu\nu}$ and the asterisk defines the conjugate fields:

$$s^{*\mu\nu} = a^{\mu\rho}\, a^{\nu\sigma}\, \underline{s}^{*}_{\rho\sigma} = \frac{+1}{2\sqrt{-a}}\, \epsilon^{\mu\nu\rho\sigma}\, s_{\rho\sigma},$$

$$\underline{s}^{*}_{\mu\nu} = \frac{-\sqrt{-a}}{2}\, \epsilon_{\mu\nu\rho\sigma}\, \underline{s}^{\rho\sigma}, \qquad (6.5)$$

Here $\epsilon^{\mu\nu\rho\sigma} = \pm 1,\ 0$ is the Levi-Civita tensor.

We consider now the scalar density

$$\mathcal{L}_B = 2\sqrt{-\pi}^{\,*} \qquad (6.6)$$

where π is the determinant of the tensor

$$\pi_{\mu\nu} = a_{\mu\nu} + s_{\mu\nu} \qquad (6.7)$$

Calculation of the determinants similar to those of Chapter I leads to

$$\pi = a + s + \frac{a}{2}\, a^{\mu\rho}\, a^{\nu\sigma} s_{\mu\nu} s_{\rho\sigma}\,, \qquad (6.8)$$

that is

$$\mathcal{L}_B = 2\sqrt{-a}\; L, \qquad (6.9)$$

where

$$L = \left(1 + \frac{s}{a} + \frac{1}{2}\, a^{\mu\rho}\, a^{\nu\sigma}\, s_{\mu\nu} s_{\rho\sigma}\right)^{\frac{1}{2}} = (1 + F - G^2)^{\frac{1}{2}}. \qquad (6.10)$$

We can now define the conjugate field from $\mathcal{L}_B$**

$$\sqrt{-a}\; p^{\mu\nu} = \frac{\partial \mathcal{L}_B}{\partial s_{\mu\nu}} = \sqrt{-a}\left(2\,\frac{\partial L}{\partial F}\, \underline{s}^{\mu\nu} + \frac{\partial L}{\partial G}\; s^{*\mu\nu}\right) \qquad (6.11)$$

or

$$p^{\mu\nu} = \frac{\underline{s}^{\mu\nu} - G\, s^{*\mu\nu}}{L}. \qquad (6.12)$$

---

*Born and Infeld do not start from the density $\mathcal{L}_B = \sqrt{-\pi}$ but from $\mathcal{L} = \sqrt{-\pi} - \sqrt{-a}$.

** The derivation of $\mathcal{L}$ and L are carried out by assuming that $s_{\mu\nu}$ and $s_{\nu\mu}$ are independent variables. (See footnote p. 20.)

A variational principle applied to $\mathcal{L}_B$ leads uniquely to Euler's equations

$$\mathfrak{I}^{\mu} = \partial_{\rho}(\sqrt{-a}\; p^{\mu\rho}) = 0. \tag{6.13}$$

A second relation is obtained by assuming arbitrarily that

$$s_{\mu\nu\rho} = \partial_{\mu} s_{\nu\rho} + \partial_{\rho} s_{\mu\nu} + \partial_{\nu} s_{\rho\mu} = 0$$

or

$$\partial_{\rho}(\sqrt{-a}\; s^{*\mu\rho}) = 0. \tag{6.14}$$

Instead of considering $\mathcal{L}_B$ as a function of $a_{\mu\nu}$ and $s_{\mu\nu}$, we can define a Hamiltonian

$$\mathcal{H}_B = 2\sqrt{-\chi}, \tag{6.15}$$

χ being the determinant of the tensor

$$\chi_{\mu\nu} = a_{\mu\nu} + p^{*}_{\mu\nu}. \tag{6.16}$$

We have then the following results:

$$\mathcal{H}_B = 2\sqrt{-a}\; H, \tag{6.17}$$

where

$$H = \sqrt{\frac{\chi}{a}} = \left(1 + \frac{p^*}{a} + \frac{1}{2} a^{\mu\nu} a^{\rho\sigma} p^{*}_{\mu\rho} p^{*}_{\nu\sigma}\right)^{\frac{1}{2}} = (1 + P - Q^2)^{\frac{1}{2}}, \tag{6.18}$$

with

$$P = \frac{1}{2} p^{*\mu\nu} p^{*}_{\mu\nu} = \frac{1}{2} a^{\mu\rho} a^{\nu\sigma} p^{*}_{\rho\sigma} p^{*}_{\mu\nu}, \tag{6.19}$$

$$Q = \frac{1}{4} p^{\mu\nu} p^{*}_{\mu\nu} = \frac{-1}{8\sqrt{-a}} \epsilon^{\mu\nu\rho\sigma} p^{*}_{\rho\sigma} p^{*}_{\mu\nu} = -\frac{p^*}{\sqrt{a}}$$

$$\text{for } p^{\mu\nu} = \frac{-1}{2\sqrt{-a}} \epsilon^{\mu\nu\rho\sigma} p^{*}_{\rho\sigma}. \tag{6.20}$$

We then define the conjugate fields $s^*_{\mu\nu}$ from $p^{*\mu\nu}$

$$\sqrt{-a}\, s^{*\mu\nu} = \frac{\partial \mathcal{H}}{\partial p^*_{\mu\nu}} = \sqrt{-a}\left(2\frac{\partial H}{\partial P}\, p^{*\mu\nu} + \frac{\partial H}{\partial Q}\, p^{\mu\nu}\right) \tag{6.21}$$

or

$$s^{*\mu\nu} = \frac{p^{*\mu\nu} - Qp^{\mu\nu}}{H}. \tag{6.22}$$

In the particular case of a static spherically symmetric solution, the non-linear relations between the displacement and the electric fields lead to the conclusion that the displacement $D(p^{14})$ tends to infinity while the electric field $E(s_{14})$ stays finite. We thus have

$$D_r = \frac{e}{r^2}, \qquad E = \frac{e}{r_0^2\left(1 + \frac{r^4}{r_0^4}\right)^{\frac{1}{2}}}. \tag{6.23}$$

Born's theory proposes two equivalent interpretations of charged particles: one interpretation considers them as point charges producing the displacement D which is discontinuous at the origin and the other assumes that the particle can be replaced by a continuous charge distribution characterized by the field E remaining finite at the origin. The latter conception leads to the definition of a free charge density ρ and a free current which are functions not of the coordinates, but of the field quantities. The current density is then defined by

$$\mathcal{J}^{\mu} = \frac{1}{4\pi}\,\partial_{\rho}(\sqrt{-a}\, s^{\mu\rho}) \tag{6.24}$$

while the four-vectors $\partial_\rho(\sqrt{-a}\, p^{\mu\rho})$ and $\partial_\rho(\sqrt{-a}\, s^{*\mu\rho})$ both vanish. In the spherically symmetric static case, integration of (6.24) leads to the charge of the particle.

C. THE NON-LINEAR RELATIONS BETWEEN THE DISPLACEMENT AND THE ELECTRIC FIELDS IN THE UNIFIED THEORY.

The interpretation of the notion of a charged particle suggested by Born's theory seems particularly adapted to the principles of a field theory such as the unified field theory. The hamiltonian in Einstein's theory is expressed in terms of the Ricci tensor $R_{\mu\nu}$ and introduces two types of tensors: on the one hand the $g^{\mu\nu}(h^{\mu\nu}, f^{\mu\nu})$ and the $g_{\mu\nu}$ $(\gamma_{\mu\nu}, \varphi_{\mu\nu})$ and on the other hand $R_{\mu\nu}$.

Thus there are four tensors $(f^{\mu\nu}, \varphi_{\mu\nu}, R_{\mu\nu}, R^{\mu\nu})$ that can be associated with the electromagnetic field and as we have seen (cf. § 2) two tensors $(\gamma_{\mu\nu}, h^{\mu\nu})$ that can be associated with the metric in a manner which is still to be specified.

In unified theory, the most common form for the hamiltonian is the homogenous density.

$$\mathfrak{H} = \sqrt{-g}\, g^{\mu\nu} R_{\mu\nu}. \qquad (2.17)$$

If we want to define conjugate fields with respect to $\mathfrak{H}$ and associate them by nonlinear expressions as given in Born's theory, it is necessary to adopt a purely affine theory by specifying conveniently $\mathfrak{H}$. In this case, the variations $\partial\Gamma^{\rho}_{\mu\nu}$ of the only independent variables lead to corresponding variations $\partial R_{\mu\nu}$ such that

$$\delta \int \mathfrak{H}\, d\tau = \int \mathcal{G}^{\mu\nu}\, \delta R_{\mu\nu}\, d\tau = 0,$$

where

$$\mathcal{G}^{\mu\nu} = \frac{\partial \mathfrak{H}}{\partial R_{\mu\nu}} \quad . \qquad (2.16)$$

Without writing explicitly the form of $\mathfrak{H}$, we obtain only the group of equations (cf. Chapter 2, § 2) which, in terms of the connection

$$L^{\rho}_{\mu\nu} = \Gamma^{\rho}_{\mu\nu} + \frac{2}{3}\delta^{\mu}_{\rho}\, \Gamma_{\nu}\ (L_{\rho} = 0)$$

can be written

$$D_\rho \mathcal{G}^{\mu\nu}_{+-} = 0, \qquad \partial_\rho \mathfrak{g}^{\mu\rho} = 0$$

The other field equations are obtained from (2. 16) and assume consequently a convenient form of the hamiltonian $\mathfrak{H}$.

In 1943, Schrödinger had tried to obtain nonlinear relations between the conductions and the fields analogous to those of Born's theory (22). To this end, he used instead of the density $\mathfrak{H}\,(R_{\mu\nu})$ the expression

$$\mathfrak{H}\,(h^{\mu\nu}, \underset{\vee}{R}_{\mu\nu}) = \sqrt{-g}\, h^{\mu\nu} \underline{R_{\mu\nu}} - \mathfrak{H}\,(R_{\mu\nu}). \qquad (6.\ 25)$$

He defined the conjugate quantities in terms of $\overline{\mathfrak{H}}$:

$$\underline{R_{\mu\nu}} = \frac{\partial\, \overline{\mathfrak{H}}}{\partial \mathfrak{H}^{\mu\nu}}, \qquad \mathfrak{g}^{\mu\nu} = \frac{\partial\, \overline{\mathfrak{H}}}{\partial \underset{\vee}{R}_{\mu\nu}}. \qquad (6.\ 26)$$

To derive the second group of the equations of the theory, Schrödinger specifies $\overline{\mathfrak{H}}\,(h^{\mu\nu}, \underset{\vee}{R}_{\mu\nu})$ in the following manner (cf. (22), p. 51)

$$\overline{\mathfrak{H}} = 2\,\alpha\{\sqrt{-\det(h_{\mu\nu} + \underset{\vee}{R}_{\mu\nu})} - \sqrt{-\det h_{\mu\nu}}\}. \qquad (6.27)$$

This definition is very close to the one Born proposes. Substitution of (6. 27) into (6. 26) leads to:"

$$\mathfrak{g}^{\mu\nu} = \frac{\alpha\sqrt{-h}}{A}\,(h^{\mu\rho}\, h^{\nu\sigma} \underset{\vee}{R}_{\rho\sigma} - I_2\, \underset{\vee}{R}^{*\mu\nu}), \qquad (6.\ 28)$$

$$\underline{R_{\mu\nu}} = \frac{\alpha}{A}[h^{\rho\sigma}\, \underset{\vee}{R}_{\mu\rho}\, \underset{\vee}{R}_{\nu\sigma} - h_{\mu\nu}(A - 1)], \qquad (6.\ 29)$$

where

$$\left.\begin{aligned} \underset{\vee}{R}^{*\mu\nu} &= \frac{1}{2\sqrt{-h}}\, \epsilon^{\mu\nu\rho\sigma}\, \underset{\vee}{R}_{\rho\sigma}, \qquad I_2 = \frac{1}{4}\, \underset{\vee}{R}^{*\mu\nu}\, \underset{\vee}{R}_{\mu\nu}, \\ A &= (1 + \frac{1}{2}\, h^{\mu\rho}\, h^{\nu\sigma}\, \underset{\vee}{R}_{\mu\nu}\, \underset{\vee}{R}_{\rho\sigma} - I_2^2)^{\frac{1}{2}}. \end{aligned}\right\} \qquad (6.\ 30)$$

Equations (6.28) form a system of non-linear relations between inductions and electromagnetic fields and permit us to express the tensor $R_{\underline{\mu\nu}} - \frac{1}{2} h_{\mu\nu} h^{\rho\sigma} R_{\underline{\rho\sigma}}$ as a function of a maxwellian energy tensor.

Schrödinger (75) has rewritten the definitions (6.28) and (6.29) in the particular case of a static spherically symmetric solution. Using polar coordinates, he seeks a solution of the type

$$h_{11} = - e^{\lambda}, \quad h_{22} = - r^2, \quad h_{33} = - r^2 \sin^2\theta, \quad h_{44} = e^{\nu};$$

$$\underset{\vee}{R_{41}} = - \underset{\vee}{R_{14}} = \psi(r),$$

λ, ν, and ψ being functions of r. The metric and the electromagnetic field have a form which is identical to the forms of Papapetrou. However, here the various quantities are associated with $h_{\mu\nu}$ and $\underset{\vee}{R_{\mu\nu}}$ (and not $\gamma_{\mu\nu}$ and $\varphi_{\mu\nu}$) which do not form the symmetric and antisymmetric parts of $g_{\mu\nu}$ (or $g^{\mu\nu}$). Integration of $\partial_\rho \mathfrak{g}^{\mu\rho} = 0$ and (6.29) leads to the solutions

$$e^{\nu} = e^{-\lambda} = 1 - \frac{f_1^2}{r} \int_0^r (\sqrt{\epsilon^2 + r^4} - r^2)\, dr,$$

$$\psi = \frac{b\epsilon}{c\sqrt{\epsilon^2 + r^4}}$$

where f_1, b, ϵ and c are constants. This is identical with the electric field proposed by Born and Infeld and consequently finite at the origin. Schrödinger also shows that for $r \gg r_0$, we have

$$e^{\nu} = e^{-\lambda} = 1 - \frac{2k}{c^2} \frac{m_0}{r}$$

instead of

$$e^{\nu} = e^{-\lambda} = 1 - \frac{2m}{r} + \frac{4\pi e^2}{r^2}$$

which result in general Relativity from Schwarzschild's solution with the electrostatic case taken into account.

This theory, while quite attractive in view of the physical interpretations, has the great inconvenience of departing considerably from the more general and more symmetrical form of Einstein's initial theory. This is essentially due to the hybrid character of the field variables, the expression of $\bar{\mathfrak{H}}$ associating $R_{\underset{\vee}{\mu\nu}}$ to $h_{\mu\nu}$: it follows therefore that the true hamiltonian $\mathfrak{H}(R_{\mu\nu})$ is a complicated expression of the functions $R_{\underline{\mu\nu}}$ and $R_{\underset{\vee}{\mu\nu}}$. In addition, (6. 26) applied to $\mathfrak{H}$ will not lead to the second group of the field equations.

$$W_{\underline{\mu\nu}} = \lambda\gamma_{\mu\nu}, \quad \partial_\rho W_{\underset{\vee}{\mu\nu}} + \partial_\nu W_{\underset{\vee}{\rho\mu}} + \partial_\mu W_{\underset{\vee}{\nu\rho}} = \lambda\varphi_{\mu\nu\rho}.$$

The relations (6. 28) and (6. 29) which replace them will evidently lead to different solutions. This will be clearer in the determination of the static spherically symmetric solution which will not be the generalized Schwarzschild solution we obtained in Chapter 5.

Nevertheless, we have seen that it is possible to maintain in a purely affine theory the general relations of Einstein's theory. It is sufficient to specify $\mathfrak{H}$ in generalizing the expression

$$\mathfrak{H}^{(0)} = \frac{2}{\lambda}\sqrt{-\det G_{\mu\nu}}$$

proposed by Weyl and Eddington where $G_{\mu\nu}(\{\ \})$ is Ricci's tensor relative to a Riemannian variety. We can define as in (2. 18) (Chapter 2) the scalar density.

$$\mathfrak{H} = \frac{2}{\lambda}\sqrt{-\det R_{\mu\nu}} \tag{2. 18}$$

as done by Schrödinger. Substitution of (2. 18) in (2. 16) leads to (cf. Chapter 2 § 2)

$$\lambda \mathcal{G}^{\mu\nu} = - \frac{\text{minor } R_{\mu\nu}}{\sqrt{-\det R_{\mu\nu}}} = - \frac{[R]\, R^{\mu\nu}}{\sqrt{-[R]}} = \sqrt{-[R]}\, R^{\mu\nu}, \tag{6. 31}$$

with

$$[R] = \det R_{\mu\nu}, \tag{6.32}$$

$$R_{\mu\nu} = \lambda\, g_{\mu\nu} \tag{6.33}$$

The resultant field equations are those occurring in the mixed theory (cf. Chapter 2) by adoption of a non-homogenous hamiltonian

$$\mathfrak{H} = \mathcal{G}^{\mu\nu}R_{\mu\nu} - 2\lambda\sqrt{-g}. \tag{6.34}$$

With (6.31), Eqs. (6.33) are the non-linear relations between the conjugate variables (with respect to $\mathfrak{H}$) $G^{\mu\nu}$ and $R_{\mu\nu}$. In fact, splitting into symmetric and antisymmetric parts

$$R_{\mu\nu} = \Gamma_{\mu\nu} + \Phi_{\mu\nu}, \tag{6.35}$$

$$R^{\mu\nu} = H^{\mu\nu} + F^{\mu\nu} \tag{6.36}$$

and letting $[R]$, Γ, Φ be the determinants of $R_{\mu\nu}$, $\Gamma_{\mu\nu}$, and $\Phi_{\mu\nu}$, we have the standard relations of determinant theory:

$$[R] = \Gamma + \Phi + \frac{\Gamma}{2}\Gamma^{\mu\rho}\,\Gamma^{\nu\sigma}\Phi_{\mu\nu}\Phi_{\rho\sigma}, \tag{6.37}$$

$$H^{\mu\nu} = \frac{\Gamma}{[R]}\,\Gamma^{\mu\nu} + \frac{\Phi}{[R]}\,\Phi^{\mu\rho}\Phi^{\nu\sigma}\Gamma_{\rho\sigma}, \tag{6.38}$$

$$F^{\mu\nu} = \frac{\Phi}{[R]}\,\Phi^{\mu\nu} + \frac{\Gamma}{[R]}\,\Gamma^{\mu\rho}\Gamma^{\nu\sigma}\Phi_{\rho\sigma}. \tag{6.39}$$

Thus (6.31) becomes

$$\lambda\sqrt{-g}\,h^{\mu\nu} = -\frac{1}{\sqrt{-[R]}}\left[\Gamma\,\Gamma^{\mu\nu} + \Phi\Phi^{\mu\rho}\Phi^{\nu\sigma}\Gamma_{\rho\sigma}\right], \tag{6.40}$$

$$\lambda\sqrt{-g}\,f^{\mu\nu} = -\frac{1}{\sqrt{-[R]}}\left[\Phi\Phi^{\mu\nu} + \Gamma\,\Gamma^{\mu\rho}\,\Gamma^{\nu\sigma}\,\Phi_{\rho\sigma}\right]. \tag{6.41}$$

It is then preferable, as Einstein believed, to eliminate all heterogeniety from the theory and consequently to exclude with the λ terms, Eqs. (6. 40) and (6. 41) which result from the preceding theory.

If we were then to start from a mixed theory without a cosmological term, the tensors $\mathbf{R}_{\mu\nu}$ and $\mathcal{G}^{\mu\nu}$ will remain conjugate with respect to the density $\mathcal{L}$. On the other hand, the

$$\underset{\vee}{W}_{\mu\nu} = \underset{\vee}{R}_{\mu\nu} + \frac{2}{3}(\partial_\mu \Gamma_\nu - \partial_\nu \Gamma_\mu)$$

and the $\mathcal{G}_{\mu\nu}$ will satisfy the equations (see Chapter 4 § 4)

$$\partial_\rho \mathfrak{I}^{\mu\rho} = 0, \qquad \partial_\rho \underset{\vee}{W}_{\mu\nu} + \partial_\nu \underset{\vee}{W}_{\rho\mu} + \partial_\mu \underset{\vee}{W}_{\nu\rho} = 0 \quad (6.42)$$

analogous to those of Born's theory. But there are then no non-linear relations between them.

These non-linear relations between $\mathcal{G}^{\mu\nu}$ and $g_{\mu\nu}$ exist by definition and it is possible to assume that the fields and the inductions are associated, in a manner to be specified, to $\varphi_{\mu\nu}(g_{\mu\nu})$ and to the tensor density $\sqrt{-g}\, f^{\mu\nu}(\mathcal{G}^{\mu\nu}_{\vee})$. However, these quantities (although conjugate with respect to $\mathcal{L} = 2\sqrt{-g}$ and associated by non-linear expressions) are not conjugate with respect to the homogenious hamiltonian of the theory. In addition, they will not satisfy the fundamental relations (6. 13) and (6. 14) of Born's theory since one of them only, $\sqrt{-g}\, f^{\mu\nu}$, has a vanishing divergence.

Thus, the interpretation of the antisymmetric fields introduced by the theory is in part uncertain.

Kursunoglu (107) has re-examined this question by associating the electromagnetic field and inductions with the tensor $\varphi_{\mu\nu}$ and its dual $\varphi^{*\mu\nu} = \left(1/2\sqrt{-\gamma}\right)\epsilon^{\mu\nu\rho\sigma}\varphi_{\rho\sigma}$. He introduces a nonhomogeneous lagrangian which is in our notation

$$\mathcal{L}_1 = \mathcal{G}^{\mu\nu} R_{\mu\nu} - 2p^2\left(\sqrt{-\gamma} - \sqrt{-g}\right) = \mathcal{G}^{\mu\nu} R_{\mu\nu} - 2p^2\sqrt{-\gamma}\left(1 - \frac{\overline{g}}{\sqrt{\gamma}}\right)^*$$

The quantity $\sqrt{g/\gamma}$ is the scalar $1 + F - G^2$ defined in Chapter 1

* Kursunoglu's b is $\frac{g^2}{h}$ in our notation and $\frac{g^2}{h} = \gamma$ (1.22)

and p is a fundamental constant which has the dimensions of a reciprocal length. This theory which starts from a new version of Einstein's theory (39) uses approximate solutions for the static spherically symmetric solution and attempts to find the following results: finite self-energy for the election, equations of motion, bending of light, etc.

Recently, Mavridès (65) has re-examined this question in trying to remove the indeterminacy of the metric and the fields. His results are discussed in the following section.

D. DEFINITION OF CONJUGATE FIELDS.

Consider a tensor which is split into symmetric and antisymmetric parts:

$$\pi_{\mu\nu} = a_{\mu\nu} + s_{\mu\nu}. \tag{6.43}$$

Let π, a, s denote the determinates of $\pi_{\mu\nu}$, $a_{\mu\nu}$ and $s_{\mu\nu}$ and let $\pi\pi^{\mu\nu}$, $aa^{\mu\nu}$ and $ss^{\mu\nu}$ be the minors relative to the corresponding elements. $\pi\,\mu\nu$ can be written in terms of its symmetric and antisymmetric parts as

$$\pi^{\mu\nu} = b^{\mu\nu} + q^{\mu\nu}. \tag{6.44}$$

The relations between $a_{\mu\nu}$, $s_{\mu\nu}$, $b^{\mu\nu}$ and $q^{\mu\nu}$ are those defined for $\gamma_{\mu\nu}$, $\varphi_{\mu\nu}$, $h^{\mu\nu}$ and $f^{\mu\nu}$ in Chapter 1. Let

$$\pi_{\mu\nu} = \mathcal{G}\, g_{\mu\nu} \begin{cases} a_{\mu\nu} = T\gamma_{\mu\nu}, \\ s_{\mu\nu} = T\varphi_{\mu\nu}, \end{cases} \tag{6.45}$$

where $\mathcal{G}$ is an invariant. We then have

$$\pi = \mathcal{G}^4\, g, \tag{6.46}$$

and

$$\pi^{\mu\nu} = \frac{1}{\pi}\text{ minor } \pi_{\mu\nu} = \frac{\mathcal{G}^3}{\mathcal{G}^4 g}\text{ minor } g_{\mu\nu} = \frac{1}{\mathcal{G}}\, g^{\mu\nu}. \tag{6.47}$$

The relations (6.47) are equivalent to

$$\pi^{\mu\nu} = \frac{1}{\mathcal{J}} g^{\mu\nu} \begin{cases} b^{\mu\nu} = \frac{1}{\mathcal{J}} h^{\mu\nu}, \\ q^{\mu\nu} = \frac{1}{\mathcal{J}} f^{\mu\nu}. \end{cases} \tag{6.48}$$

a. Let us assume that the true metric is $a_{\mu\nu}$ and one of the conjugate fields is $s_{\mu\nu}$. Then

$$a = \mathcal{J}^4 \gamma, \quad a^{\mu\nu} = \frac{1}{a} \text{ minor } a_{\mu\nu} = \frac{1}{\mathcal{J}} \gamma^{\mu\nu} \tag{6.49}$$

and we can define the quantities

$$s^{\underline{\mu\nu}} = a^{\mu\rho} a^{\nu\sigma} s_{\rho\sigma} = \frac{1}{\mathcal{J}} \gamma^{\mu\rho}\gamma^{\nu\sigma} \varphi_{\rho\sigma}, \tag{6.50}$$

$$s^{*\mu\nu} = \frac{1}{2\sqrt{-a}} \epsilon^{\mu\nu\rho\sigma} s_{\rho\sigma} = \frac{1}{2\,\mathcal{J}\sqrt{-\gamma}} \epsilon^{\mu\nu\rho\sigma} \varphi_{\rho\sigma}, \tag{6.51}$$

and the two invariants

$$F = \frac{1}{2} s_{\mu\nu} s^{\underline{\mu\nu}} = \frac{1}{2} \gamma^{\mu\rho} \gamma^{\nu\sigma} \varphi_{\mu\nu} \varphi_{\rho\sigma}, \tag{6.52}$$

$$G = \frac{1}{4} s_{\mu\nu} s^{*\mu\nu} = \frac{1}{8\sqrt{-\gamma}} \epsilon^{\mu\nu\rho\sigma} \varphi_{\mu\nu} \varphi_{\rho\sigma} = \sqrt{\frac{\varphi}{-\gamma}}. \tag{6.53}$$

If we consider the invariant density $\mathcal{L}(\gamma, \varphi)$ defined in (1. 27):

$$\mathcal{L} = 2\sqrt{-g} = 2\left[\left(\gamma + \varphi + \frac{\gamma}{2} \gamma^{\mu\rho}\gamma^{\nu\sigma} \varphi_{\mu\nu} \varphi_{\rho\sigma}\right)\right]^{\frac{1}{2}}, \tag{6.54}$$

that is

$$\mathcal{L} = 2\sqrt{-\gamma}\, L, \tag{6.55}$$

where

$$L = \sqrt{\frac{g}{\gamma}} = \left(1 + \frac{1}{2} \gamma^{\mu\nu} \gamma^{\rho\sigma} \varphi_{\mu\rho} \varphi_{\nu\sigma} + \frac{\varphi}{\gamma}\right)^{\frac{1}{2}} = (1 + F - G^2)^{\frac{1}{2}} \tag{6.56}$$

The relations (cf. footnote p. 17)

$$f^{\mu\nu} = \frac{1}{\sqrt{-g}} \frac{\partial \mathcal{L}}{\partial \varphi_{\mu\nu}} = \frac{\varphi}{g} \varphi^{\mu\nu} + \frac{\gamma}{g} \gamma^{\mu\rho} \gamma^{\nu\sigma} \varphi_{\rho\sigma}, \tag{6.57}$$

can then be written

$$2\frac{\partial L}{\partial s_{\mu\nu}} = \frac{s^{\underline{\mu\nu}} - G s^{*\mu\nu}}{(1 + F - G^2)^{\frac{1}{2}}} = \frac{1}{\mathcal{G}}\sqrt{\frac{\gamma}{g}}\left(\gamma^{\mu\rho}\gamma^{\nu\sigma}\varphi_{\rho\sigma} + \frac{\varphi}{\gamma}\varphi^{\mu\nu}\right)$$

$$= \frac{1}{\mathcal{G}}\sqrt{\frac{g}{\gamma}}\, f^{\mu\nu} \tag{6.58}$$

or

$$\frac{s^{\underline{\mu\nu}} - G s^{*\mu\nu}}{L} = p^{\mu\nu}, \tag{6.59}$$

where

$$p^{\mu\nu} = \frac{1}{\mathcal{G}}\sqrt{\frac{g}{\gamma}}\, f^{\mu\nu}. \tag{6.60}$$

b. Let us now assume that the true metric $b_{\mu\nu}$ and the fields are associated with $g^{\mu\nu}$. Let

$$\pi^{\mu\nu} = \frac{1}{\mathcal{G}} g^{\mu\nu} \begin{cases} b^{\mu\nu} = \frac{1}{\mathcal{G}} h^{\mu\nu}, \\ q^{\mu\nu} = \frac{1}{\mathcal{G}} f^{\mu\nu}. \end{cases} \tag{6.61}$$

The determinant b of $b_{\mu\nu}$ will then be

$$b = \mathcal{G}^4\, h, \qquad b_{\mu\nu} = \mathcal{G}\, h_{\mu\nu}. \tag{6.62}$$

We then have

$$q_{\underline{\mu\nu}} = b_{\mu\rho}\, b_{\nu\sigma}\, q^{\rho\sigma} = \mathcal{G}\, h_{\mu\rho}\, h_{\nu\sigma}\, f^{\rho\sigma}, \tag{6.63}$$

$$q^*_{\mu\nu} = \frac{-\sqrt{-b}}{2}\,\epsilon_{\mu\nu\rho\sigma}\, q^{\rho\sigma} = \frac{-\mathcal{G}\sqrt{-h}}{2}\,\epsilon_{\mu\nu\rho\sigma}\, f^{\rho\sigma} \tag{6.64}$$

and the following expressions of the two invariants

$$F = \frac{1}{2} q_{\underline{\mu\nu}}\, q^{\mu\nu} = \frac{1}{2} h_{\mu\rho}\, h_{\nu\sigma}\, f^{\mu\nu} f^{\rho\sigma}$$

$$\left(= \frac{1}{2}\gamma^{\mu\rho}\gamma^{\nu\sigma}\varphi_{\mu\nu}\varphi_{\rho\sigma}\right), \tag{6.65}$$

$$G = \frac{1}{4} q^{*\underline{\mu\nu}} q_{\underline{\mu\nu}} = -\frac{\sqrt{-h}}{8} \epsilon_{\mu\nu\rho\sigma} f^{\mu\nu} f^{\rho\sigma} = -\sqrt{\frac{-h}{f}}$$

$$\left(= \sqrt{\frac{\varphi}{-\gamma}}\right). \tag{6. 66}$$

Let us consider the function $\mathcal{L}(h_{\mu\nu}, f_{\mu\nu})$ defined in (1. 28)

$$\mathcal{L} = 2\sqrt{-g} = 2\sqrt{-h}\left(1 + \frac{h}{f} + \frac{1}{2} h_{\mu\rho} h_{\nu\sigma} f^{\mu\nu} f^{\rho\sigma}\right)^{-\frac{1}{2}} \tag{6. 67}$$

that is

$$\mathcal{L} = \frac{2\sqrt{-h}}{L} \tag{6. 68}$$

with

$$L = \sqrt{\frac{g}{\gamma}} = \sqrt{\frac{h}{g}} = \left(1 + \frac{1}{2} h_{\mu\rho} h_{\nu\sigma} f^{\mu\nu} f^{\rho\sigma} + \frac{h}{f}\right)^{\frac{1}{2}} = (1 + F - G^2)^{\frac{1}{2}}.$$

$$\tag{6. 69}$$

The relations

$$-\frac{1}{\sqrt{-g}} \frac{\partial \mathcal{L}}{\partial f^{\mu\nu}} = \frac{2}{L} \frac{\partial L}{\partial f^{\mu\nu}} = \frac{g}{f} f_{\mu\nu} + \frac{g}{h} h_{\mu\rho} h_{\nu\sigma} f^{\rho\sigma} = \varphi_{\mu\nu}$$

can then be written (see footnote p. 21)

$$2\frac{\partial L}{\partial q_{\underline{\mu\nu}}} = \frac{q^{\mu\nu} - G q^{\underline{\mu\nu}*}}{(1 + F - G^2)^{\frac{1}{2}}}$$

$$= \frac{1}{\mathcal{G}} \sqrt{\frac{g}{h}} \left(f^{\mu\nu} + \frac{h}{f} h^{\mu\rho} h^{\nu\sigma} f_{\rho\sigma}\right)$$

$$= \frac{1}{\mathcal{G}} \sqrt{\frac{h}{g}} h^{\mu\rho} h^{\nu\sigma} \varphi_{\rho\sigma} \tag{6. 70}$$

or

$$\frac{q^{\mu\nu} - G q^{\underline{\mu\nu}*}}{L} = r^{\underline{\mu\nu}}, \tag{6. 71}$$

where

$$r_{\mu\nu} = \mathcal{G} \sqrt{\frac{h}{g}} \varphi_{\mu\nu}. \tag{6. 72}$$

Choice of the invariant. It is natural to assume that the densities formed from the field $p^{\mu\nu}$ and the metric a—or from the field $q^{\mu\nu}$ and the metric b—will satisfy the usual divergence relations. However the relation

$$\partial_\rho \mathfrak{g}^{\mu\rho} = \partial_\rho(\sqrt{-g}\, f^{\mu\rho}) = 0$$

can be written in the form

$$\begin{cases} \text{with the choice } a_{\mu\nu}\text{: } \partial_\rho(\mathcal{G}\sqrt{-\gamma}\, p^{\mu\rho}) = \partial_\rho\left(\dfrac{\sqrt{-a}}{\mathcal{G}} p^{\mu\rho}\right) = 0; \\ \text{with the choice } b^{\mu\nu}\text{: } \partial_\rho(\sqrt{-g}\, \mathcal{G}\, q^{\mu\rho})\left(= \partial_\rho \dfrac{\sqrt{-g}}{\mathcal{G}\sqrt{-h}} \sqrt{-b}\, q^{\mu\rho}\right) = 0. \end{cases} \tag{6.73}$$

We will then have

$$\partial_\rho \mathcal{P}^{\mu\rho} = 0 \text{ if } \mathcal{G} = 1 \text{ in the case a } (\mathcal{P}^{\mu\rho} = \sqrt{-a}\, p^{\mu\rho}), \tag{6.74}$$

$$\partial_\rho \mathcal{Q}^{\mu\rho} = 0 \text{ if } \mathcal{G} = \sqrt{\frac{g}{h}} = \sqrt{\frac{\gamma}{g}} \text{ in the case b } (\mathcal{Q}^{\mu\rho} = \sqrt{-b}\, q^{\mu\rho}). \tag{6.75}$$

The metric and the fields will then be defined in the following way

$$\begin{cases} \text{a. } a_{\mu\nu} = \gamma_{\mu\nu}, & \text{b. } \begin{cases} b^{\mu\nu} = \sqrt{\dfrac{g}{\gamma}}\, h^{\mu\nu}, \\ \text{or } \sqrt{-b}\, b^{\mu\nu} = \sqrt{-g}\, h^{\mu\nu}, \end{cases} \\ s_{\mu\nu} = \varphi_{\mu\nu}, & r_{\mu\nu} = \varphi_{\mu\nu}, \\ p^{\mu\nu} = \sqrt{\dfrac{g}{\gamma}}\, f^{\mu\nu}; & q^{\mu\nu} = \sqrt{\dfrac{g}{\gamma}}\, f^{\mu\nu}. \end{cases} \tag{6.76}$$

Thus, in both cases, the fields and inductions can be expressed in the same way as a function of $\varphi_{\mu\nu}$ and $f^{\mu\nu}$. On the other hand, the metric $a_{\mu\nu}$ and $b_{\mu\nu}$ are such that

$$a = b = \gamma.$$

Application to the case of spherical symmetry. Let us assume that a static spherically symmetric solution has only a single antisymmetric field φ_{14} so that we can use Papapetrou's solution (5. 52) since this is the only usable solution. We then have (by putting V = 1 in (5. 52):

$$g_{\mu\nu} = \begin{pmatrix} -\alpha & & & w \\ & -\beta & & \\ & & -\beta\sin^2\theta & \\ -w & & & \sigma \end{pmatrix}, \quad \text{with} \begin{cases} \alpha = \dfrac{1}{1 - \dfrac{2m}{r}}, \\ \sigma = \left(1 - \dfrac{2m}{r}\right)\left(1 + \dfrac{k^2}{r^4}\right), \\ \beta = r^2, \quad w = \dfrac{k}{r^2}, \end{cases} \tag{6.77}$$

that is

$$g = -(\alpha\sigma - w^2)\beta^2 \sin^2\theta = -r^4 \sin^2\theta, \tag{6.78}$$

$$\gamma = -\alpha\sigma\beta^2 \sin^2\theta = -\left(1 + \frac{k^2}{r^4}\right) r^4 \sin^2\theta, \tag{6.79}$$

$$\sqrt{\frac{g}{\gamma}} = \frac{1}{\sqrt{1 + \dfrac{k^2}{r^4}}}$$

Using (1. 18), we can deduce the components $g^{\mu\nu}$

$$h^{\mu\nu} = \frac{\gamma}{g}\gamma^{\mu\nu} + \frac{\varphi}{g}\varphi^{\mu a}\varphi^{\nu b}\gamma_{ab},$$

$$f^{\mu\nu} = \frac{\varphi}{g}\varphi^{\mu\nu} + \frac{\gamma}{g}\gamma^{\mu a}\gamma^{\nu b}\varphi_{ab} \tag{6. 80}$$

that is

$$g^{\mu\nu} = \begin{pmatrix} -\sigma & & & -w \\ & -\dfrac{1}{\beta} & & \\ & & -\dfrac{1}{\beta\sin^2\theta} & \\ w & & & \alpha \end{pmatrix}$$

$$f^{14} = \frac{\gamma}{g}\gamma^{11}\gamma^{44}\varphi_{14} = -\frac{k}{r^2} \tag{6. 81}$$

Depending on whether we choose a or b, the metric tensor will have the components:

$$\begin{cases} \text{a. Choice } a_{\mu\nu} = \gamma_{\mu\nu} & \text{b. Choice } b^{\mu\nu} = \dfrac{g}{\gamma} h^{\mu\nu} \\ & \left(\text{or } b_{\mu\nu} = \dfrac{\gamma}{g} h_{\mu\nu}\right). \\ a_{11} = \dfrac{-1}{1 - \dfrac{2m}{r}}, & b_{44} = -\dfrac{1}{b_{11}} = \left(1 - \dfrac{2m}{r}\right) \quad 1 + \dfrac{k^2}{r^4}, \\ a_{22} = \dfrac{a_{33}}{\sin^2\theta} = -r^2, & b_{22} = \dfrac{b_{33}}{\sin^2\theta} = -r^2 \quad 1 + \dfrac{k^2}{r^4}. \\ a_{44} = \left(1 - \dfrac{2m}{r}\right)\left(1 + \dfrac{k^2}{r^4}\right). & \end{cases} \tag{6.82}$$

and the components of the electromagnetic inductions and the field will be the same in (a) as in (b).

$$s_{14} = r_{14} = \varphi_{14} = \frac{k}{r^2}, \tag{6.83}$$

$$p^{14} = q^{14} = \frac{g}{\gamma} f^{14} = -\frac{k}{r^2} \frac{1}{1 + \dfrac{k^2}{r^4}}. \tag{6.84}$$

These expressions can be rewritten in the form

$$b\,s_{14} = \frac{e}{r^2} = D, \tag{6.85}$$

$$-bp^{14} = \frac{e}{r^2} \frac{1}{1 + \dfrac{r_0^4}{r^4}} = E, \tag{6.86}$$

where

$$kb = e, \qquad k = r_0^2. \tag{6.87}$$

While the induction D becomes infinite as r tends to zero, the field E remains finite at the origin. These conclusions are those predicted by Born's theory. The value of E—

defined here in terms of its contravariant components $p^{\mu\nu}$— due to a generalization ($w \neq 0$) of Schwarzschild's solution, is the one predicted in Born's theory.

E. CURRENT AND CHARGE.

a. Let us assume that we have chosen the metric $a_{\mu\nu} = \gamma_{\mu\nu}$, and reconsider the equation

$$\mathcal{G}^{\mu} = \partial_{\rho}(\sqrt{-g}\, f^{\mu\rho}) = 0,$$

that is

$$\partial_{\rho}\,(\sqrt{-a}\, p^{\mu\rho}) = 0. \tag{6.88}$$

From (6. 58), we have

$$\partial_{\rho}(\sqrt{-a}\; p^{\mu\rho}) = \partial_{\rho}\;\left[\sqrt{-a}\;\left(2\,\frac{\partial L}{\partial F}\, s^{\underline{\mu\rho}} + \frac{\partial L}{\partial G}\, s^{*\mu\rho}\right)\right] = 0 \tag{6.89}$$

or

$$4\pi\sqrt{-a}\; j^{\mu} = \partial_{\rho}\;(\sqrt{-a}\; s^{\underline{\mu\rho}}), \tag{6.90}$$

where

$$-4\pi\, j^{\mu} = \frac{1}{2\dfrac{\partial L}{\partial F}}\;\left[2\, s^{\underline{\mu\rho}}\;\partial_{\rho}\left(\frac{\partial L}{\partial F}\right) + s^{*\mu\rho}\;\partial_{\rho}\left(\frac{\partial L}{\partial G}\right) \right. \tag{6.91}$$

$$\left. + \frac{1}{\sqrt{-a}}\;\frac{\partial L}{\partial G}\;\partial_{\rho}(\sqrt{-a}\; s^{*\mu\rho})\right]\;.$$

The term $\partial_{\rho}(\sqrt{-a}\; s^{*\mu\rho}) = \frac{1}{6}\,\epsilon^{\mu\nu\lambda\tau}(\partial_{\nu}s_{\lambda\tau} + \partial_{\lambda}s_{\tau\nu} + \partial_{\tau}s_{\nu\lambda})$ brings in the pseudo-vector $\varphi_{\mu\nu\rho} = \partial_{\rho}\varphi_{\mu\nu}$ and does not vanish when we maintain the usual assumptions of the unified theory which do not assume that $\varphi_{\mu\nu\rho} = 0$. In the particular case of a static spherically symmetric solution with only φ_{14} ($\varphi_{\mu\nu\rho} = 0$), this term does not enter. In this case j^{μ} as defined by (6. 90) reduces to j^{4} such that

$$4\pi j^4 = \frac{1}{\sqrt{-a}}\,\partial_1(\sqrt{-a}\,s^{\underline{4}\,\underline{1}})\,. \tag{6.92}$$

But from (6.82) and (6.83)

$$s^{\underline{4}\,\underline{1}} = a^{44}\,a^{11}\,s_{41} = \frac{k}{r^2}\;\frac{1}{1+\frac{k^2}{r^4}}\,, \tag{6.93}$$

$$\sqrt{-a} = \sqrt{-} \;= r^2\,\sin\theta \quad 1+\frac{k^2}{r^4}\,. \tag{6.94}$$

Substituting in (6.92), we have

$$4\pi j^4 = \frac{1}{r^2\,\sqrt{1+\frac{k^2}{r^4}}}\;\frac{\partial}{\partial r}\left(\frac{k}{\sqrt{1+\frac{k^2}{r^4}}}\right) = \frac{2k^3}{r^7\,\sqrt{1+\frac{k^2}{r^4}}\left(1+\frac{k^2}{r^4}\right)^{\frac{3}{2}}}$$

$$\frac{2}{r\,\sqrt{1+\frac{k^2}{r^4}}\left(1+\frac{r^4}{k^2}\right)^{\frac{3}{2}}}\,. \tag{6.95}$$

If we let as in (6.87)

$$k\,b = e, \qquad k = r_0^2\,, \tag{6.96}$$

then (6.95) can be written

$$b\,j^4 = \frac{e}{2\pi r_0^3\left(\frac{r}{r_0}\right)\left(1+\frac{r^4}{r_0^4}\right)^{\frac{3}{2}}}\;\frac{1}{\sqrt{1+\frac{k^2}{r^4}}}\,. \tag{6.97}$$

Integrating (6.97) over all space we have

$$\int \rho\,dv = \int_0^{\infty}\int_0^{\pi}\int_0^{2\pi} b\,j^4\sqrt{-a}\,dv, \quad \text{with } dv = d\varphi\,d\theta\,dr \tag{6.98}$$

or

$$\int \rho \, dv = \int_0^\infty \int_0^\pi \int_0^{2\pi} \frac{er^2 \sin\theta}{2\pi r_0^3 \left(\frac{r}{r_0}\right)\left(1 + \frac{r^4}{r_0^4}\right)^{\frac{3}{2}}} \, d\varphi \, d\theta \, dr \tag{6.99}$$

$$= \frac{2e}{r_0^3} \int_0^\infty \frac{r^2 \, dr}{\frac{r}{r_0}\left(1 + \frac{r^4}{r_0^4}\right)^{\frac{3}{2}}} = e \int_0^{\frac{\pi}{2}} \cos\psi \, d\psi = e,$$

where

$$\mathrm{tg}\,\psi = \frac{r^2}{r_0^2}.$$

Thus, the charge distribution, whose expression as a function of the fields is characterized by the free density j^4, leads to a finite expression when integrated over all space. In this case with $a_{\mu\nu}$ as the metric, we can associate p^{14} and s_{14} or $\sqrt{g/\gamma}\; f^{14}$ and φ_{14} with the electric field.

b. The expression (2. 27)

$$\mathcal{G}^{\mu} = \partial_\rho \, \mathcal{G}^{\mu\rho} = 0$$

cannot lead to a valid definition of the current in case b. In fact, (2. 27) can be written as

$$\partial_\rho \left[\frac{\partial \mathcal{L}(\varphi, \gamma)}{\partial \varphi_{\mu\rho}}\right] = \partial_\rho \left[\sqrt{-\gamma} \; \frac{\partial L(\varphi, \gamma)}{\partial \varphi_{\mu\rho}}\right] = 0 \tag{6.100}$$

and leads us to define a current $\partial_\rho(\sqrt{-\gamma}\; \gamma^{\mu\lambda}\gamma^{\rho\sigma}\varphi_{\lambda\sigma})$ $= \partial_\rho(\sqrt{-a}\, s^{\mu\rho})$ deduced from $L(\varphi, \gamma)$. This current is the divergence of a field $s_{\lambda\sigma} = \varphi_{\lambda\sigma}$ associated with the metric $\gamma_{\mu\nu}$ $= a_{\mu\nu}$ which permits the definition of the contravariant components $\varphi^{\underline{\mu}\,\underline{\nu}} = \gamma^{\mu\lambda}\, \gamma^{\nu\sigma}\, \varphi_{\lambda\sigma}$. This definition is not valid in the case b of the metric $b^{\mu\nu} = h^{\mu\nu}$.

To obtain a satisfactory definition of the current, we start from the pseudo-vector

$$\varphi_{\mu\nu\rho} = \partial_\rho \varphi_{\mu\nu} + \partial_\nu \, \varphi_{\rho\mu} + \partial_\mu \, \varphi_{\nu\rho}$$

which, in general, does not vanish in unified theory. We then have using (6. 76b) and (6. 70)

$$\varphi_{\mu\nu\rho} = \partial_\rho r_{\mu\nu} + \partial_\nu r_{\rho\mu} + \partial_\mu r_{\nu\rho}, \quad (6.101)$$

with

$$r_{\mu\nu} = 2 b_{\mu\rho} b_{\nu\sigma} \frac{\partial L}{\partial q_{\underline{\rho\sigma}}} = 2 \frac{\partial L}{\partial F} q_{\underline{\mu\nu}} + \frac{\partial L}{\partial G} q^*_{\mu\nu}, \quad (6.102)$$

L (f, h) being expressed by (6. 69). Substitution into (6.101) leads to

$$\varphi_{\mu\nu\rho} = \left(2 \frac{\partial L}{\partial F} \partial_\rho q_{\underline{\mu\nu}} + 2 q_{\underline{\mu\nu}} \partial_\rho \frac{\partial L}{\partial F} + q^*_{\mu\nu} \partial_\rho \frac{\partial L}{\partial G}\right) \quad (6.103)$$

$$+ \text{circular permutation of } \mu, \nu, \rho.$$

But from (6. 73)

$$\frac{1}{2} \epsilon^{\mu\nu\sigma\tau} \partial_\nu q^*_{\sigma\tau} = \partial_\rho (\sqrt{-b}\, q^{\mu\rho}) = 0 \quad (6.104)$$

Hence (6. 103) becomes

$$-4\pi \sqrt{-b}\, i^\mu = \partial_\rho (\sqrt{-b}\, q^{*\underline{\mu\rho}})$$

$$= \frac{1}{6} \epsilon^{\mu\rho\lambda\sigma} (\partial_\rho q_{\underline{\lambda\sigma}} + \partial_\sigma q_{\underline{\rho\lambda}} + \partial_\lambda q_{\underline{\sigma\rho}}), \quad (6.105)$$

where

$$-4\pi i^\rho = \frac{1}{2 \frac{\partial L}{\partial F}} \left[2 q^{*\underline{\mu\rho}} \partial_\rho \left(\frac{\partial L}{\partial F}\right. + q^{\mu\rho} \partial_\rho \frac{\partial L}{\partial G} + \partial_\rho (\sqrt{-b}\, r^{*\mu\rho})\right]. \quad (6.106)$$

We can immediately see the similarity between (6. 106) and (6. 91) but, in case b, the current i^μ is the divergence of the field $q^{*\mu\rho}$. In the particular case of spherical symmetry, the definition of the charge density $\sqrt{-b}\, i^4$ is associated with $\partial_1 \sqrt{-b}\, q^{*41}$ that is the component

$$q_{\underline{23}} = b_{22} b_{33} q^{23} = \sqrt{\frac{\gamma}{g}}\, h_{22} h_{33} f^{23}. \quad (6.107)$$

This assumes a satisfactory and possible definition of the induction f^{23} and the field φ_{23}. But, if it seems preferable to associate the electric field to the components φ_{pq} * (p. q = 1, 2, 3) and the metric to $h_{\mu\nu}$**, it neverthless remains true that Wyman's solution in φ_{23} does not lend to physically acceptable results in the limit of the so-called "strong conditions." It is possible that this problem cannot be solved without associating to φ_{14} the field φ_{23} in the spherically symmetric case.† In any case, if we adopt the definition (6. 101) for the current, we must, in the spherically symmetric case, examine all the possibilities discussed in Chapter 5.††

If we limit ourselves to a spherically symmetric solution associated with the antisymetric field φ_{14} and to the choice $a_{\mu\nu} = \gamma_{\mu\nu}$ for the metric, it is possible at the present stage of the theory, to define a field which remains finite at the origin and to interpret the charges as a spatial continuous distribution whose free density is only a function of a field everywhere finite. Thus, we come close to a conception of charges an example of which is given by Born's theory in a limited domain. This conception seems necessary to a pure field theory as Einstein's unified theory proposes to be.

---

* cf. § 2, Chapter 4.

** cf. § 1, Chapter 6 and Lichnerowicz ((10), p. 288).

† As carried out by Bonnor for the strong system (§ 12, Chapter 5).

†† cf. § 14, Chapter 5.

7

Some Problems Raised by the Unified Field Theory

A. ENERGY-MOMENTUM TENSOR.

1. ENERGY MOMENTUM TENSOR IN GENERAL RELATIVITY AND THE CONSERVATION EQUATIONS.

In General Relativity, the effects of the gravitational field described by Einstein's tensor $T_{\mu\nu}$ are balanced by the energy-momentum tensor $S_{\mu\nu}$. While $S_{\mu\nu}$ is expressed uniquely as a function of the geometrical properties of the universe, $T_{\mu\nu}$ has purely phenomenological origin. The fundamental equation relative to the interior case

$$S_{\mu\nu} = G_{\mu\nu} - \frac{1}{2} g_{\mu\nu} G = \chi T_{\mu\nu}$$

leads to the identity

$$\nabla_\rho T_\mu^\rho = 0$$

or

$$\partial_\rho \mathfrak{T}_\mu{}^\rho = \left\{ {\sigma \atop \mu\rho} \right\} \mathfrak{T}_\sigma{}^\rho = \frac{1}{2} \mathfrak{T}^{\sigma\rho} \partial_\mu g_{\sigma\rho}, \tag{7.1}$$

where $\mathfrak{T}_\mu^\rho = \sqrt{-g}\, T_\mu{}^\rho$.

In any coordinate system, the energy density $\mathfrak{T}_\mu^\rho$ does not satisfy a conservation equation. But, we can define a pseudodensity t_μ^ρ which, added to $\mathfrak{T}_\mu^\rho$, satisfies the divergence equation (cf. (5), p. 135).

$$\partial_\rho(\mathfrak{T}_\mu^{\ \rho} + t_\mu^{\ \rho}) = 0. \tag{7.2}$$

From (1.2), we have

$$\begin{aligned} G_{\mu\nu}\, d\mathfrak{g}^{\mu\nu} &= \sqrt{-g}(-G^{\mu\nu}\, dg_{\mu\nu} + \frac{1}{2} G\, g^{\mu\nu}\, dg_{\mu\nu}) \\ &= -(G^{\mu\nu} - \frac{1}{2} g^{\mu\nu} G)\sqrt{-g}\, dg_{\mu\nu} = -\chi \mathfrak{T}^{\mu\nu}\, dg_{\mu\nu}. \end{aligned} \tag{7.3}$$

But, we always have the relation

$$G_{\mu\nu} = \partial_\rho \frac{\partial \mathfrak{A}}{\partial(\partial_\rho \mathfrak{g}^{\mu\nu})} - \frac{\partial \mathfrak{A}}{\partial \mathfrak{g}^{\mu\nu}}, \tag{7.4}$$

where

$$\mathfrak{A} = \mathfrak{g}^{\mu\nu}\left[\left\{{\rho \atop \mu\nu}\right\}\left\{{\lambda \atop \rho\lambda}\right\} - \left\{{\rho \atop \mu\lambda}\right\}\left\{{\lambda \atop \nu\rho}\right\}\right]^{*}. \tag{7.5}$$

Substitution of (7.4) into (7.3) leads to

$$-\chi\mathfrak{T}^{\mu\nu}\partial_\sigma g_{\mu\nu} = \partial_\rho\left\{(\partial_\sigma \mathfrak{g}^{\mu\nu})\frac{\partial \mathfrak{A}}{\partial(\partial_\rho \mathfrak{g}^{\mu\nu})} - \delta_\sigma^\rho A\right\}, \tag{7.6}$$

or, if we use (7.1)

$$\partial_\rho \mathfrak{T}_\sigma^{\ \rho} = -\partial_\rho t_\sigma^{\ \rho}, \tag{7.7}$$

where

$$2\chi\, t_\sigma^{\ \rho} = (\partial_\sigma \mathfrak{g}^{\mu\nu})\frac{\partial \mathfrak{A}}{\partial(\partial_\rho \mathfrak{g}^{\mu\nu})} - \delta_\sigma^\rho \mathfrak{A}. \tag{7.8}$$

We will then always have (7.2) by the introduction of t_μ^ρ which is not a tensor density** and can be made to correspond to a potential energy by analogy with classical mechanics.

---

* This is the reason that the basic equation $G_{\mu\nu} = 0$ which is equivalent to the $\delta\int \mathfrak{A}\, d\tau = 0$ (7.5) can be deduced from a variational principle.

** t_μ^ρ can always be made to vanish at a point by a covariant choice of the coordinates (cf. e.g., Eddington (5), p. 136).

2. THE ENERGY TENSOR IN THE UNIFIED THEORY. CONSERVATION RELATIONS.

We have established the conservation relations (2. 61) (cf. Chapter 2, § C) (115), (116), (41) in the unified field theory:

$$\partial_\rho \mathcal{K}_\mu^{\ \rho} + \frac{1}{2} W_{\lambda\tau}\partial_\mu \mathcal{G}^{\lambda\tau} = 0,$$

where

$$K_\mu^{\ \rho} = \frac{1}{2}(W_{\mu\nu}\, g^{\rho\nu} + W_{\nu\mu} g^{\nu\rho}) - \frac{1}{2}\, \delta_\mu^{\ \rho}\, W \quad (W = g^{\mu\nu} W_{\mu\nu}). \qquad (2.56)$$

$W_{\mu\nu}$ being Ricci's tensor (2. 36) formed with the connection $L^\rho_{\mu\nu}$, whose torsion vanishes.

Schrödinger has shown that these identities can be written in a form which generalizes (7. 2) (115). He makes use of the development given in Chapter 2 without making the assumption that the ξ^ρ which define the transformation vanish at the integration limits. We shall follow Schrödinger's method as developed in (115), (116), (117).

We consider the expression

$$\Lambda = \mathcal{G}^{\mu\nu}\, \Lambda_{\mu\nu}, \qquad (7.9)$$

with

$$\Lambda_{\mu\nu} = L^\sigma_{\mu\rho} L^\rho_{\sigma\nu} - L^\rho_{\mu\nu} L^\sigma_{\rho\sigma}, \quad L^{\ \sigma}_{\underset{\vee}{\rho\sigma}} = 0. \qquad (7.10)$$

It can be shown that

$$\Lambda^\rho_{\mu\nu}\, \delta(\partial_\rho \mathcal{G}^{\mu\nu}) = -2\Lambda_{\mu\nu}\, \delta\mathcal{G}^{\mu\nu} - \mathcal{G}^{\mu\nu}\, \delta\Lambda_{\mu\nu}, \qquad (7.11)$$

where

$$\Lambda^{\ \rho}_{\mu\nu} = L^\rho_{\mu\nu} - \delta^\rho_\nu L^\rho_{\mu\sigma}. \qquad (7.12)$$

We then have from (7. 9) and (7. 11)

$$\delta\Lambda = -\Lambda_{\mu\nu}\, \delta\mathcal{G}^{\mu\nu} - \Lambda^\rho_{\mu\nu}\, \delta(\partial_\rho \mathcal{G}^{\mu\nu}) \qquad (7.13)$$

and, for the variations $\delta \mathcal{G}^{\mu\nu}$ which do not necessarily vanish at the limits of integrations

$$\delta \int \Lambda \, d\tau = \int (\partial_\rho \Lambda^\rho_{\mu\nu} - \Lambda_{\mu\nu}) \, \delta \mathcal{G}^{\mu\nu} d\tau - \int \partial_\rho (\Lambda^\rho_{\mu\nu} \, \delta \mathcal{G}^{\mu\nu}) \, d\tau$$

$$= \int W_{\mu\nu} \, \delta \mathcal{G}^{\mu\nu} \, d\tau - \int \partial_\rho (\Lambda^\rho_{\mu\nu} \, \delta \mathcal{G}^{\mu\nu}) \, d\tau$$

$$= \chi \int \mathfrak{T}_{\mu\nu} \, \delta g^{\mu\nu} \, d\tau - \int \partial_\rho (\Lambda^\rho_{\mu\nu} \, \delta \mathcal{G}^{\mu\nu}) \, d\tau. \tag{7.14}$$

In fact, we have as in (7. 3)

$$W_{\mu\nu} \, \delta \mathcal{G}^{\mu\nu} = W_{\mu\nu} \, \sqrt{-g} (\delta g^{\mu\nu} - \frac{1}{2} g^{\mu\nu} \, g_{\rho\sigma} \, \delta g^{\rho\sigma}) \tag{7.15}$$

$$= \sqrt{-g} \, (W_{\mu\nu} - \frac{1}{2} \, g_{\mu\nu} \, W) \, \delta g^{\mu\nu} = \chi \mathfrak{T}_{\mu\nu} \, \delta g^{\mu\nu},$$

where

$$W_{\mu\nu} - \frac{1}{2} \, g_{\mu\nu} \, W = \chi \, T_{\mu\nu}, \qquad \mathfrak{T}_{\mu\nu} = \sqrt{-g} \, T_{\mu\nu}$$

and from (2. 60)

$$\chi \mathfrak{T}^\nu_\mu = \frac{\chi}{2} \, (\mathfrak{T}_{\mu\rho} \, g^{\nu\rho} + \mathfrak{T}_{\rho\mu} g^{\rho\nu}) = \mathcal{K}^\nu_\mu .$$

We identify now $\delta g^{\mu\nu}$ with the Lie derivative obtained for the infinitesimal charge of coordinates

$$x'^\rho = x^\rho + \xi^\rho,$$

and obtain (cf. footnote p. 43)

$$\delta \, g^{\mu\nu} = g^{\mu\rho} \, \frac{\partial \xi^\nu}{\partial x^\rho} + g^{\rho\nu} \, \frac{\partial \xi^\mu}{\partial x^\rho} - \frac{\partial g^{\mu\nu}}{\partial x^\rho} \, \xi^\rho,$$

$$\delta \mathcal{G}^{\mu\nu} = \mathcal{G}^{\mu\rho} \, \frac{\partial \xi^\nu}{\partial x^\rho} + \mathcal{G}^{\rho\nu} \, \frac{\partial \xi^\mu}{\partial x^\rho} - \frac{\partial}{\partial x^\rho} (\mathcal{G}^{\mu\nu} \xi^\rho) \tag{7.16}$$

The corresponding variation of $\int \Lambda \, d\tau$ is then

$$\delta \int \Lambda \, d\tau = \int \delta \Lambda \, d\tau + \int \partial_\rho (\Lambda \xi^\rho) \, d\tau \tag{7.17}$$

where the last integral

$$\int_{D'} \Lambda(x')\, d\tau' - \int_{D} \Lambda(x)\, d\tau$$

arises from the variation of the limits of integration. If we assume that the variations $\delta \mathcal{G}^{\mu\nu}$ which occur in (7.13) can be expressed in terms of (7.16) we can replace $\delta\Lambda$ in (7.17) by its value (7.13) (See Schrödinger (115)) and obtain (7.18)

$$\delta \int \Lambda\, d\tau = \chi \int \mathfrak{T}_{\mu\nu}\, \delta g^{\mu\nu}\, d\tau \qquad (7.18)$$

$$+ \int \partial_\rho (\Lambda \xi^\rho - \Lambda^{\rho}_{\mu\nu}\, \delta \mathcal{G}^{\mu\nu})\, d\tau.$$

Using (7.16) and integrating by parts, we have

$$\delta \int \Lambda\, d\tau = -2 \int \left[\partial_\sigma \mathcal{K}^{\sigma}_{\rho} + \frac{\chi}{2} \mathfrak{T}_{\mu\nu}\, \partial_\rho g^{\mu\nu}\right] \xi^\rho\, d\tau \qquad (7.19)$$

$$+ \int \partial_\rho [\Lambda \xi^\rho - \Lambda^{\rho}_{\mu\nu}\, \delta \mathcal{G}^{\mu\nu} + 2\, \mathcal{K}^{\rho}_{\sigma}\, \xi^\sigma]\, d\tau.$$

If we take into account (7.15) and the identity (2.61)

$$\partial_\rho\, K_{\mu}{}^{\rho} + \frac{1}{2} W_{\alpha\beta}\, \partial_\mu\, \mathcal{G}^{\alpha\beta} = 0.$$

derived in Chapter 2, then the first integral on the right-hand side of (7.19) will vanish. Substitution of $\delta \mathcal{G}^{\mu\nu}$ in the second integral by its value (7.16) leads to:

$$\delta \int \Lambda\, d\tau = \int \partial_\sigma \{(\delta^{\sigma}_{\rho} \Lambda + \Lambda^{\sigma}_{\mu\nu}\, \partial_\rho \mathcal{G}^{\mu\nu} + 2\chi \mathfrak{T}^{\sigma}_{\rho}) \xi^\rho \qquad (7.20)$$

$$-(\Lambda^{\sigma}_{\rho\nu} \mathcal{G}^{\lambda\nu} + \Lambda^{\sigma}_{\nu\rho} \mathcal{G}^{\nu\lambda} - \delta^{\lambda}_{\rho}\, \Lambda^{\sigma}_{\mu\nu}\, \mathcal{G}^{\mu\nu})\, \partial_\lambda \xi^\rho\}\, d\tau.$$

However, in a linear transformation, the components of the affine connection transform as a tensor and consequently Λ behaves as a scalar density. (7.20) must be invariant independently of the ξ^ρ. In particular if the ξ^ρ are constant, (7.20) reduces to

$$\partial_\sigma(\mathfrak{T}_\rho^{\ \sigma} + t_\rho^{\sigma}) = 0, \tag{7.21}$$

with

$$\chi\, t_\rho^{\sigma} = \frac{1}{2}\,(\delta_\rho^{\sigma}\Lambda + \Lambda_{\mu\nu}^{\ \sigma}\,\partial_\rho\, \mathcal{G}^{\mu\nu}). \tag{7.22}$$

This is exactly eq. (7.2). On the other hand, substitution of (7.21) into (7.20) leads to

$$\mathfrak{T}_\rho^{\sigma} + t_\rho^{\sigma} = \frac{1}{2}\,\partial_\mu(\Lambda_{\rho\nu}^{\mu}\,\mathcal{G}^{\sigma\nu} + \Lambda_{\nu\rho}^{\mu}\,\mathcal{G}^{\nu\sigma} - \delta_\rho^{\sigma}\Lambda_{\lambda\nu}^{\ \mu}\,\mathcal{G}^{\lambda\nu}) \tag{7.23}$$

where we have taken (7.22) into account and assumed that $\delta_\sigma\,\xi^\rho$ are arbitrary constants.

Papapetrou has established relations, analogous to (7.21), for the tensors $\mathcal{S}_\rho^{\ \sigma}$ and s_ρ^σ (112) [$\mathcal{S}_\rho^{\ \sigma}$ being expressed as a function of $R_{\mu\nu}$ just as $\mathfrak{T}_\rho^\sigma$ is a function of $W_{\mu\nu}$ and $s_\rho^\sigma = t_\rho^\sigma - \frac{2}{3\chi}(\partial_\rho\,\mathcal{G}^{\sigma\lambda})\,\Gamma_\lambda$]. He then shows that for sufficiently weak static and spherical fields to admit the approximation

$$g_{ik} \cong \epsilon_{ik} + \mathcal{E}'_{ik}, \quad \text{against}\ \mathcal{E}'_{ik} \sim \frac{1}{r}, \tag{7.24}$$

we can write

$$m = \int (\mathfrak{T}_4^{\ 4} - \mathfrak{T}_1^1 - \mathfrak{T}_2^2 - \mathfrak{T}_3^3)\; dv, \tag{7.25}$$

m being the mass which occurs in Schwarzschild solution. For a static field such that $T_{\mu\nu} = 0$, we must have $m = 0$. This generalizes for the unified theory the theorem proven by Einstein and Pauli for general relativity (9). All static non-singular fields ($T_{\mu\nu} = 0$) correspond to a vanishing total mass (cf. (112) p. 1108).

3. THE TENSOR $\Psi_\mu^{\ \nu}$ IN UNIFIED FIELD THEORY.

According to Einstein, an essential principle of the unified field theory is to develop a complete geometrical description of the unified field. Thus, there is only one case that is unified and the tensor $T_{\mu\nu}$ must be extracted from the first set of the field equations. Referring to the field equations (4. 3), we have for the symmetrical part

$$G_{\mu\nu} = \Phi_{\mu\nu}, \tag{7. 26}$$

with

$$\Phi_{\mu\nu} = \nabla_\rho u^\rho_{\mu\nu} - \frac{1}{2}\nabla_\mu \nabla_\nu \operatorname{Log} g + u^\rho_{\mu\nu}\, u_\rho - u^\lambda_{\mu\rho} u^\rho_{\lambda\nu} - \underset{v}{L}{}^\lambda_{\mu\rho}\, \underset{v}{L}{}^\rho_{\lambda\nu}\,. \tag{7.27}$$

These equations are rigorous. $G_{\mu\nu}$ is Ricci's tensor formed with the Christoffel symbols $\left\{ {\rho \atop \mu\nu} \right\}$ relative to $\gamma_{\mu\nu}$ and satisfies the relation

$$\nabla_\rho \left(G^\rho_\mu - \frac{1}{2}\,\delta^\rho_\mu\, G \right) = 0,$$

where V_ρ is the covariant derivative in terms of the Christoffel symbols and the indices of $G_{\mu\nu}$ are raised with the help of γ. We must therefore show that $\Phi_{\mu\nu}$ satisfies the identity

$$\nabla_\rho (\Phi^\rho_\mu - \frac{1}{2}\,\delta^\rho_\mu\, \Phi) = 0. \tag{7. 28}$$

This equation simplifies considerably if we use the approximate values obtained by assuming that $\varphi_{\mu\nu}$ is small and of the order ($\cong \epsilon$). There the $\gamma_{\mu\nu}$ differ from the galilean values to order ϵ^2. We thus obtain (4. 44).

$$\underset{2}{\mathcal{L}}_{\mu\nu} + \underset{2}{Q}_{\mu\nu} = 0.$$

Using the notation of (4.60), we can write

$$\underset{2}{Q}_{\mu\nu} = \Phi_{\mu\nu} = -\epsilon_\rho\epsilon_\sigma \Big(\frac{1}{4}\,\partial_\mu\partial_\nu\, \underset{1}{\varphi}_{\rho\sigma}\underset{1}{\varphi}_{\rho\sigma} - \partial_\sigma \underset{1}{\varphi}_{\rho\nu}\partial_\rho\, \underset{1}{\varphi}_{\mu\sigma} + \frac{1}{4}\,\underset{1}{\varphi}_{\sigma\nu\rho}\underset{1}{\varphi}_{\sigma\mu\rho} - \frac{1}{2}\,\underset{1}{\varphi}_{\nu\sigma}\,\partial_\rho \underset{1}{\varphi}_{\sigma\mu\rho} - \frac{1}{2}\,\underset{1}{\varphi}_{\mu\sigma}\partial_\rho \underset{1}{\varphi}_{\sigma\nu\rho} \Big) \tag{4. 64}$$

and from (4. 22), we have

$$\underset{2}{\mathcal{L}}_{\mu\nu} = \frac{1}{2}\epsilon_\rho(\partial_\rho\partial_\mu \underset{2}{\gamma}_{\nu\rho} - \partial_\rho\partial_\rho\underset{2}{\gamma}_{\mu\nu} - \partial_\nu\partial_\mu\underset{2}{\gamma}_{\rho\rho} + \partial_\nu\partial_\rho\underset{2}{\gamma}_{\mu\rho}) \tag{7.29}$$

The above is valid in the weak system. In the strong system, the first order equation

$$\epsilon_\rho \partial_\rho\partial_\rho\underset{1}{\varphi}_{\mu\nu\lambda} = 0 \tag{4.62}$$

can be replaced by stronger equations (cf. footnote p.79):

$$\epsilon_\rho\partial_\rho\partial_\rho \underset{1}{\varphi}_{\mu\nu} = 0, \qquad \epsilon_\rho\partial_\rho\underset{1}{\varphi}_{\mu\nu\rho} = 0, \tag{4.61}$$

and thus the last two terms of $\Phi_{\mu\nu}$ vanish.

Schrödinger shows that, in the strong system, the divergence of

$$\Psi_\mu^\rho = \Phi_\mu^\rho - \frac{1}{2}\delta_\mu^\rho\Phi \qquad (\Phi_\mu^\rho = \gamma^{\rho\sigma}\Phi_{\mu\sigma}, \Phi = \gamma^{\rho\sigma}\Phi_{\rho\sigma}) \tag{7.30}$$

vanishes identically ((117) p. 13 and 22). One can show that this holds also for the weak system (by writing $\Phi_{\mu\nu}$ in the form (4. 64) and using (4. 18) and (4. 46)).

Finally, Schrödinger seeks to simplify (7. 26)

$$G_{\mu\nu} = -\Phi_{\mu\nu} \tag{7.31}$$

by choosing a convenient coordinate transformation

$$x_\mu = x'_\mu + a_\mu(x'_\rho) \tag{7.32}$$

whose coefficients a_μ are of the second order ((117) p. 14). With this assumption, we have

$$\underset{2}{\gamma'}_{\mu\nu} = \underset{2}{\gamma}_{\mu\nu} + \epsilon_\nu\partial_\mu a_\nu - \epsilon_\mu\partial_\nu a_\mu. \tag{7.33}$$

Choosing a_μ such that

$$\epsilon_\mu\left(\partial_\mu\underset{2}{\gamma}_{\mu\nu} - \frac{1}{2}\partial_\nu\underset{2}{\gamma}_{\mu\mu}\right) + \epsilon_\mu\epsilon_\nu\partial_\mu\partial_\mu a_\nu = 0, \tag{7.34}$$

we have

$$\epsilon_\mu \left(\partial_\mu \underset{2}{\gamma'}_{\mu\nu} - \frac{1}{2}\partial_\nu \underset{2}{\gamma'}_{\mu\mu}\right) = 0. \tag{7. 35}$$

Substitution of (7. 33) into (7. 29) leads to the usual expression

$$\underset{2}{\mathcal{L}}_{\mu\nu} = \underset{2}{G}_{\mu\nu} = -\frac{\epsilon_\rho}{2}\partial_\rho\partial_\rho \underset{2}{\gamma'}_{\mu\nu} \tag{7. 36}$$

$\varphi_{\mu\nu}$ and consequently $\Phi_{\mu\nu}$ remain unchanged. We then have

$$\underset{2}{G}_{\mu\nu} - \frac{1}{2} g_{\mu\nu} \underset{2}{G} \cong -\frac{1}{2}\epsilon_\rho\partial_\rho\partial_\rho \left(\underset{2}{\gamma'}_{\mu\nu} - \frac{1}{2}\delta_{\mu\nu}\epsilon_\mu\epsilon_\sigma\underset{2}{\gamma'}_{\sigma\sigma}\right) = -\underset{2}{\Psi}_{\mu\nu}. \tag{7. 37}$$

The expression $\underset{2}{\gamma'}_{\mu\nu} - \frac{1}{2}\delta_{\mu\nu}\epsilon_\mu\epsilon_\sigma\underset{2}{\gamma'}_{\sigma\sigma}$ corresponds to the retarded potentials of $\Psi_{\mu\nu}$ whose divergence vanishes. The divergence of this expression therefore vanishes. This condition is precisely (7. 35). Thus the retarded potentials corresponding to (7. 37) constitute an approximate solution of the rigorous equations (7. 26). This solution is not unique.

4. THE PSEUDODENSITY $t_{\mu\rho}$ IN UNIFIED THEORY.

The symmetric tensor $\Psi_{\mu\nu}$ which occurs on the right-hand side of (7. 37) could be identified with the energy tensor whose properties it has. However, according to Schrödinger ((117), p. 23), there are two other tensors that fulfill these conditons.

The first is the pseudodensity, up to a function

$$t_\sigma^\rho = \frac{1}{2}[(\partial_\sigma \mathcal{G}^{\mu\nu})\Lambda_{\mu\nu}^\rho + \delta_\sigma^\rho \Lambda], \tag{7. 38}$$

where Λ, $\Lambda_{\mu\nu}^\rho$ are defined by (7. 9), (7. 10) and (7. 12). The other is a second pseudodensity

$$\underset{A}{t}_\sigma^\rho = -\frac{1}{2}\mathcal{G}^{\mu\nu}\partial_\sigma \Lambda_{\mu\nu}^\rho . \tag{' 39}$$

We can immediately see that t_σ^ρ generalizes the pseudodensity defined in (7. 8) and, up to a sign,* reduces to it when $L_{\mu\nu}^\rho = \{^{\rho}_{\mu\nu}\}$. On the other hand t_σ^ρ does not reduce to (7. 8) even when $L_{\mu\nu}^\rho = \{^{\rho}_{\mu\nu}\}$. Schrödinger associates the existence of this tensor to the purely affine aspect of the theory ((117), p. 24). We can easily check that both pseudodensities have a vanishing divergence in the approximation considered

$$\partial_\rho t_\sigma^\rho = 0, \qquad \partial_\rho \underset{A}{t_\sigma^\rho} = 0. \tag{7. 40}$$

Finally, Schrödinger has computed the values of t_σ^ρ and $\underset{A}{t_\sigma^\rho}$ for weak fields. In the above approximation and the Schrödinger notation (4. 60) we have (see (117) Eqs. (3. 9), (3.10))

$$\begin{aligned} t_{\mu\nu} &= \frac{1}{2}\,\epsilon_\rho\epsilon_\sigma\left(\partial_\mu \underset{1}{\varphi}_{\rho\sigma} - \frac{1}{2}\underset{1}{\varphi}_{\rho\sigma\mu}\right)\partial_\nu \underset{1}{\varphi}_{\rho\sigma} \\ &- \frac{1}{8}\,\epsilon_\mu\delta_{\mu\nu}\epsilon_\rho\epsilon_\sigma\epsilon_\lambda\left[\partial_\lambda\partial_\lambda(\underset{1}{\varphi}_{\rho\sigma}\underset{1}{\varphi}_{\rho\sigma}) - \frac{1}{3}\underset{1}{\varphi}_{\rho\sigma\lambda}\underset{1}{\varphi}_{\rho\sigma\lambda}\right] \end{aligned} \tag{7. 41}$$

and

$$\begin{aligned} \underset{A}{t}_{\mu\nu} &= -\frac{1}{2}\,\epsilon_\rho\epsilon_\sigma\underset{1}{\varphi}_{\rho\sigma}\partial_\nu\left(\partial_\mu \underset{1}{\varphi}_{\rho\sigma} - \frac{1}{2}\underset{1}{\varphi}_{\rho\sigma\mu}\right) \\ &- \frac{1}{2}\,\epsilon_\rho\partial_\nu(\partial_\rho\underset{2}{\gamma}_{\mu\rho} - \partial_\mu\underset{2}{\gamma}_{\rho\rho}) + \epsilon_\rho\epsilon_\sigma\partial_\nu(\underset{1}{\varphi}_{\rho\sigma}\partial_\sigma\underset{1}{\varphi}_{\rho\mu}) \\ &- \frac{1}{2}\,\epsilon_\rho\epsilon_\sigma\partial_\nu(\underset{1}{\varphi}_{\rho\sigma}\underset{1}{\varphi}_{\rho\mu\sigma}) + \frac{1}{8}\,\epsilon_\rho\epsilon_\sigma\partial_\mu\partial_\nu(\underset{1}{\varphi}_{\rho\sigma}\underset{1}{\varphi}_{\rho\sigma}) \end{aligned} \tag{7. 42}$$

This is for the strong system to which Schrödinger refers. Choosing a reference system such that

$$\begin{aligned} \epsilon_\rho(\partial_\mu\underset{2}{\gamma}_{\rho\rho} - \partial_\rho\underset{2}{\gamma}_{\mu\rho}) + \frac{1}{2}\,\epsilon_\rho\epsilon_\sigma\Big[\partial_\mu(\varphi_{\rho\sigma}\varphi_{\rho\sigma}) \\ + \frac{1}{2}\,\varphi_{\rho\sigma}\varphi_{\rho\sigma\mu}\Big] = 0. \end{aligned} \tag{7. 43}$$

---

* We recall that from the definition of $R_{\mu\nu} = R_{\mu\nu\rho}^{\rho}$ chosen in (1. 60), the expression of $R_{\mu\nu}$ as a function of $\Gamma_{\rho\sigma}^{\lambda}$ is the one chosen by Einstein but has the opposite sign of the one adopted by Schrödinger.

(7. 42) takes the form

$$\underset{A}{t}_{\mu\nu} = \frac{1}{2}\epsilon_\rho\epsilon_\sigma \Big[\partial_\nu\varphi_{\rho\sigma}\partial_\mu\varphi_{\rho\sigma} - \frac{1}{4}\partial_\mu\partial_\nu(\varphi_{\rho\sigma}\varphi_{\rho\sigma}) - \frac{1}{2}\varphi_{\rho\sigma\mu}\partial_\nu\varphi_{\rho\sigma} + \frac{1}{4}\partial_\nu(\varphi_{\rho\sigma}\varphi_{\rho\sigma\mu})\Big] \quad (7.44)$$

The three expressions (4.64), (7. 41) and (7. 44) are different even in the order of approximation we are using. The conclusions are the same for the weak system.

B. GEODESICS AND EQUATIONS OF MOTION.

5. DEFINITION OF GEODESICS.

We shall approach the problem of defining geodesics from Eisenhart's remarks ((7) pp. 12-14). This definition results from the character of parallelism of two vectors in a non-symmetrical affine connection.

Consider a curve (C) defined by x = x(t). We shall say that the vectors $\vec{\xi}$ are parallel relative to (C) and the connection Γ if

$$\xi^\sigma\left(\frac{d\xi^\rho}{dt} + \Gamma^{\rho}_{\mu\nu}\xi^\mu\frac{dx^\nu}{dt}\right) - \xi^\rho\left(\frac{d\xi^\sigma}{dt} + \Gamma^{\sigma}_{\mu\nu}\xi^\mu\frac{dx^\nu}{dt}\right) = 0. \quad (7.45)$$

If the vector ξ^ρ represents the direction $u^\rho = \frac{dx^\rho}{dt}$ tangent to (C), these tangents will be parallel with respect to (C): the preceding equations will then represent the geodesics of the space. We thus have

$$u^\sigma\left(\frac{du^\rho}{dt} + \Gamma^{\rho}_{\mu\nu}u^\mu u^\nu\right) - u^\rho\left(\frac{du^\sigma}{dt} + \Gamma^{\sigma}_{\mu\nu}u^\mu u^\nu\right) = 0. \quad (7.46)$$

We can easily verify that, for a connection Γ' such that

$$\Gamma'^{\rho}_{\mu\nu} = \Gamma^{\rho}_{\mu\nu} + \delta^{\rho}_{\mu}\psi_\nu , \quad (7.47)$$

where ψ_ν is an arbitrary four-vector, the parallelism remains the same for (C) (cf. (7), p. 31; (22), p. 285, (10), p. 249). In particular the geodesics with respect to

$$L^{\rho}_{\mu\nu} = \Gamma^{\rho}_{\mu\nu} + \frac{2}{3}\,\delta^{\rho}_{\mu}\,\Gamma_{\nu}, \qquad (L_\rho = 0)$$

coincide with those of Γ. We shall then seek the solutions of

$$u^{\sigma}\left(\frac{du^{\rho}}{dt} + L^{\rho}_{\mu\nu}\,u^{\mu}u^{\nu}\right) - u^{\rho}\left(\frac{du^{\sigma}}{dt} + L^{\sigma}_{\mu\nu}\,u^{\mu}u^{\nu}\right) = 0. \tag{7.48}$$

We consider in particular the geodesic

$$\frac{d^2x^{\rho}}{dt^2} + L^{\rho}_{\underline{\mu\nu}}\,\frac{dx^{\mu}}{dt}\,\frac{dx^{\nu}}{dt} = \varphi(t)\,\frac{dx^{\rho}}{dt} \tag{7.49}$$

corresponding to a given function $\varphi(t)$. We can write

$$\frac{d^2x^{\rho}}{ds^2} + L^{\rho}_{\underline{\mu\nu}}\,\frac{dx^{\mu}}{ds}\,\frac{dx^{\nu}}{ds} = 0 \tag{7.50}$$

by making the substitution

$$\frac{ds}{dt} = c\,e^{\int\varphi\,dt} \tag{7.51}$$

Eq. (7.50) means that the absolute derivative of $u^{\mu} = \dfrac{dx^{\mu}}{ds}$ vanishes. In fact (118) the absolute derivative of a vector A^{μ} which can have three forms ($\overset{\mu}{A^{+}}$, $\overset{\mu}{A^{-}}$, $\overset{\mu}{A^{0}}$) reduces to the unique expression $\overset{\mu}{A^{0}}$ for u^{μ}.

Let us now consider a symmetric tensor $a_{\mu\nu}$ which is a function of g_ρ and $\Gamma^{\lambda}_{\rho\sigma}$ (Wyman (85), p. 433).

$$a_{\mu\nu} = l_{\mu\nu}\,(g_{\rho\sigma},\ \Gamma^{\lambda}_{\rho\sigma}), \tag{6.1}$$

Multiplying by $a_{\rho\lambda}\dfrac{dx^{\lambda}}{ds}$ and summing over μ, we have

$$a_{\rho\lambda}\,\frac{dx^{\lambda}}{ds}\,\frac{d^2x^{\rho}}{ds^2} + L^{\rho}_{\underline{\mu\nu}}\,a_{\rho\lambda}\,\frac{dx^{\mu}}{ds}\,\frac{dx^{\nu}}{ds}\,\frac{dx^{\lambda}}{ds} = 0 \tag{7.52}$$

or

$$\frac{d}{ds}\left(a_{\rho\lambda}\frac{dx^\rho}{ds}\frac{dx^\lambda}{ds}\right) - a_{\mu\lambda|\sigma}\frac{dx^\mu}{ds}\frac{dx^\lambda}{ds}\frac{dx^\sigma}{ds} = 0, \qquad (7.53)$$

where $a_{\mu\lambda|\sigma}$ is the covariant derivative of $a_{\mu\lambda}$ with respect to $\underline{L}^{\rho}_{\mu\nu}$. The first integral of this equation is

$$a_{\mu\lambda}\frac{dx^\mu}{ds}\frac{dx^\lambda}{ds} = \text{const.} \qquad (7.54)$$

if

$$a_{\mu\lambda|\sigma} + a_{\sigma\mu|\lambda} + a_{\lambda\sigma|\mu} = 0. \qquad (7.55)$$

a. If we assume that $a_{\mu\nu} = \gamma_{\mu\nu} (= g_{\underline{\mu\nu}})$, (7.55) is automatically satisfied. From ((S_1), p. 50) and (3.17), we have

$$a_{\mu\nu|\rho} = -(\varphi_{\mu\lambda}\underset{V}{L^{\lambda}_{\nu\rho}} + \varphi_{\nu\lambda}\underset{V}{L^{\lambda}_{\mu\rho}}) = u_{\mu\nu,\rho} \qquad (7.56)$$

(the connection Δ of S_1 being L).

Thus (7.55) can be written

$$u_{\mu\nu,\rho} + u_{\rho\mu,\nu} + u_{\nu\rho,\mu} = 0^{*} \qquad (7.57)$$

This is Eq. (3.20) which necessarily follows from the equations ((S) p. 39) of the theory. As noted by Wyman, Eq. (7.50) for the geodesic is in good agreement with the choice $ds^2 = \gamma_{\mu\nu}\, dx^\mu\, dx^\nu$ but does not require it.

b. If we define the geodesic as the shortest distance between two points

$$\delta \int ds = 0, \qquad (7.58)$$

with

$$ds^2 = a_{\mu\nu}\, dx^\mu dx^\nu, \qquad (7.59)$$

---

* We recall again that $u_{\mu\nu,\rho} = \gamma_{\rho\sigma}\, u^{\sigma}_{\mu\nu}$ and that the comma does not indicate an ordinary derivative.

we find that they depend only on the symmetrical part of the generalized metric. A development similar to the one used in general Relativity leads to

$$\frac{d^2 x^\rho}{ds^2} + \left\{ {\rho \atop \mu\nu} \right\}_{(a)} \frac{dx^\mu}{ds} \frac{dx^\nu}{ds} = 0, \qquad (7.60)$$

where $\left\{ {\rho \atop \mu\nu} \right\}_{(a)}$ are the Christoffel symbols constructed with the $a_{\mu\nu}$. The curves resulting from (7. 58) do not coincide with (7. 46) unless the condition

$$\Gamma^{\mu}_{\underline{\sigma\rho}} = \left\{ {\rho \atop \mu\nu} \right\}_{(a)} + (\delta^\rho_\mu \Psi_\nu + \delta^\rho_\nu \Psi_\mu), \qquad (7.61)$$

is satisfied. Ψ_μ is an arbitrary vector ((7), p. 31).. By definition, we have

$$L^{\rho}_{\underline{\mu\nu}} = \left\{ {\rho \atop \mu\nu} \right\} + u^{\rho}_{\mu\nu} = \Gamma^{\rho}_{\mu\nu} + \frac{1}{3}(\delta^\rho_\mu \Gamma_\nu + \delta^\rho_\nu \Gamma_\mu), \qquad (7.62)$$

where $\left\{ {\rho \atop \mu\nu} \right\}$ are the Christoffel symbols constructed from the $\gamma_{\mu\nu}$. In particular, if we choose $a_{\mu\nu} = \gamma_{\mu\nu} (= g_{\underline{\mu\nu}})$, (7. 61) reduces to

$$u^{\rho}_{\mu\nu} = -\gamma^{\rho\lambda}(\varphi_{\mu\sigma} \underset{\vee}{L^{\sigma}_{\nu\lambda}} + \varphi_\nu \underset{\vee}{L^{\sigma}_{\mu\lambda}}) = \delta^\rho_\mu \Psi'_\nu + \delta_\nu \Psi'_\mu$$

$$\left(\Psi'_\mu = \Psi_\mu + \frac{1}{3}\Gamma_\mu\right). \qquad (7.63)$$

and not to (7. 57).

6. EQUATIONS OF MOTION.

In general Relativity, the field equations, coupled with the Bianchi identities, lead to the equations of motion of a neutral particle considered as a point singularity (88), (91).

One would think that the unified field theory should allow us, in a similar way, to derive the equations of motion of a charged particle in an electromagnetic field (Lorentz equation). Infeld has remarked that one could not obtain these equations starting from the strong system. One can only deduce the equation of motion of a neutral particle as in general Relativity (101), (103).

Callaway (87) has sought to derive the equations of motion of a charged particle starting with the latest version of Einstein's theory. In the following, we shall follow closely his derivation.

We consider the quasi-static approximation studied by Einstein and Infeld (94): the derivatives with respect to time are of a lower order of magnitude than the spatial derivatives. Limiting ourselves to weak fields, we can write

$$\left.\begin{aligned}
\gamma_{44} &= 1 + \epsilon^2 \underset{2}{\gamma_{44}} + \epsilon^4 \underset{4}{\gamma_{44}} + \dots, \\
\gamma_{4\rho} &= \epsilon^3 \underset{3}{\gamma_{44}} + \epsilon^5 \underset{5}{\gamma_{44}} + \dots, \\
\gamma_{pq} &= - \delta_{pq} + \epsilon^2 \underset{2}{\gamma_{mn}} + \epsilon^4 \underset{4}{\gamma_{mn}} + \dots, \\
\varphi_{p4} &= \epsilon^3 \underset{3}{\varphi_{p4}} + \epsilon^5 \underset{5}{\varphi_{p4}}, \\
\varphi_{pq} &= \epsilon^2 \underset{2}{\varphi_{pq}} + \epsilon^4 \underset{4}{\varphi_{pq}},
\end{aligned}\right\} \qquad (7.64)$$

On the other hand, the quantities

$$\underset{\mathrm{V}}{\mathrm{L}^{\rho}_{\mu\nu}},\ \mathrm{u}^{\rho}_{\underline{\mu\nu}} = - \gamma^{\rho\sigma} (\varphi_{\mu\lambda} \underset{\mathrm{V}}{\mathrm{L}^{\lambda}_{\nu\sigma}} + \varphi_{\nu\lambda} \underset{\mathrm{V}}{\mathrm{L}^{\lambda}_{\mu\sigma}})$$

can be written in the form

$$\left.\begin{aligned}
&\underset{3}{u}^{4}_{44} = 0, \quad \underset{2}{u}^{p}_{44} = \underset{2}{u}^{4}_{p4} = \underset{2}{L}^{4}_{\underset{V}{p4}} = 0,\\
&\underset{3}{L}^{p}_{\underset{V}{4q}} = -\partial_p \underset{5}{\varphi}_{4q} + \underset{5}{\varphi}_{4pq}, \quad \underset{3}{L}^{4}_{\underset{V}{pq}} = \partial_4 \underset{3}{\varphi}_{pq} - \underset{3}{\varphi}_{pq4};\\
&\underset{2}{L}^{p}_{\underset{V}{qr}} = -\partial_p \underset{2}{\varphi}_{qr} + \underset{2}{\varphi}_{qrp},\\
&\underset{4}{L}^{4}_{\underset{V}{4p}} = \underset{4}{u}^{4}_{4p} = \underset{4}{u}^{p}_{44} = 0,
\end{aligned}\right\} \qquad (7.65)$$

$$\underset{4}{u}^{r}_{pq} = -\underset{2}{\varphi}_{pm}(-\partial_m \underset{2}{\varphi}_{qr} + \underset{2}{\varphi}_{qrm}) - \underset{2}{\varphi}_{mq}(-\partial_m \underset{2}{\varphi}_{rp} + \underset{2}{\varphi}_{rpm}),$$

$$\underset{4}{L}^{r}_{\underset{V}{pq}} = -\partial_r \underset{4}{\varphi}_{pq} + \underset{4}{\varphi}_{pqr} + \underset{2}{h}^{rs}(\partial_s \underset{2}{\varphi}_{pq} - \underset{2}{\varphi}_{pqs})$$
$$+ \underset{2}{\varphi}_{is} \left\{ {s \atop rq} \right\} - \underset{2}{\varphi}_{ks} \left\{ {s \atop ir} \right\}.$$

Substituting (7.65) into (4.46) with $\underset{2}{\varphi}_{\mu\nu\lambda}$ instead of $\underset{1}{\varphi}_{\mu\nu\lambda}$), we obtain the following equations to second and third order.

$$\nabla^2 (\underset{2}{\varphi}_{\mu\nu,\rho} + \underset{2}{\varphi}_{\rho\mu,\nu} + \underset{2}{\varphi}_{\nu\rho,\mu}) = 0, \qquad (7.66)$$

$$\nabla^2 (\underset{3}{\varphi}_{\mu 4,\rho} + \underset{3}{\varphi}_{\rho\mu,4} + \underset{3}{\varphi}_{4\rho,\mu}) = 0. \qquad (7.67)$$

In a similar way, $\partial_\rho \mathfrak{g}^{\mu\rho} = 0$ (with the approximations (4.18)) leads to

$$\operatorname{div} \underset{2}{\varphi}_{pq} = \partial_p \underset{2}{\varphi}_{pq} = 0, \qquad (7.68)$$

$$\operatorname{div} \underset{3}{\varphi}_{4p} = \partial_p \underset{3}{\varphi}_{p4} = 0, \qquad (7.69)$$

If we consider now an electric charge to be a singularity creating the field $\frac{1}{r^2}$, we must introduce the potential

$$\underset{2}{\varphi}(k) = \frac{e(k)}{r(k)}, \tag{7.70}$$

r(k) being the distance from the kth particle to the point under consideration. If the φ_{pq} are associated with the electric field and the φ_{p4} to the magnetic field, we will then have

$$\underset{2}{\varphi}_{pq} = \epsilon_{pqr}\, \partial_r\, \underset{2}{\varphi}. \tag{7.71}$$

where $\epsilon_{pqr} = 0, \pm 1$ is a totally antisymmetric tensor. This choice satisfies the second order equations (7.66) and (7.68). To obtain a solution of the third order equations (7.67) and (7.69), we assume that the motion of the kth particle is determined by three functions of ξ^m (k, t), the velocities $\dot{\xi}^m$ (k, t) being of order λ and the accelerations of order λ^2. We then have

$$\varphi_{k4} = -\epsilon_{k4pq}\, \partial_q \sum_{k=2} \varphi_2(k)\, \xi_p\, (k, t). \tag{7.72}$$

To determine the equations of motion, we must use the equations

$$\underline{W_{\mu\nu}} = 0, \qquad \partial_\rho \underset{V}{W}_{\mu\nu} + \partial_\nu \underset{V}{W}_{\rho\mu} + \partial_\mu \underset{V}{W}_{\nu\rho} = 0.$$

The second set of equations allows us to define $\underset{V}{W}_{\mu\nu}$ as the rotation of an arbitrary vector. (In fact, if we look at the derivation of the field equations from the variational principle, this arbitrary vector is just the torsion vector). The integral of $\underset{V}{W}_{\mu\nu}$ over a closed surface vanishes. Thus, the only quantities that determine the equations of motion are the $\underset{V}{W}_{\mu\nu}$. However, one can show that the equations of motion of charges result from the integration of

$$-\underset{4}{\theta}_{pq} = (\underset{4}{R}_{pq} - \underset{4}{G}_{pq}) - \frac{1}{2}\,\delta_{pq}\,\Big[(\underset{4}{R}_{ss} - \underset{4}{G}_{ss}) - (\underset{4}{R}_{44} - \underset{4}{G}_{44})\Big], \tag{7.73}$$

where G_{pq} is the Ricci tensor constructed with $\gamma_{\mu\nu}$. Substitution of (7. 67) into (7. 69) leads to (see (87), p. 1570)

$$-\underset{4}{\theta}_{pq} = \partial_r(\underset{2}{\varphi}_{ps}\,\partial_s\underset{2}{\varphi}_{qr} + \delta_{rp}\,\partial_{sq}\underset{2}{\varphi}\,\partial_s\underset{2}{\varphi} - \delta_{pq}\partial_{rs}\underset{2}{\varphi}\,\partial_s\underset{2}{\varphi}), \tag{7. 74}$$

whence

$$\partial_q\underset{4}{\theta}_{pq} = 0. \tag{7. 75}$$

A surface integral of $\underset{4}{\theta}_{pq}$ will thus be independent of the surface by virtue of (7. 75). If we choose a surface surrounding the singularity and of dimensions such that the solutions (7. 70) and (7. 71) are physically acceptable, the integral of (7. 73) will depend only on the position and the velocity of the singularity. Let n_k be the normal to the surface surrounding the kth singularity. The integral

$$\int \underset{4}{\theta}_{pk}\, n_k\, ds = 4\pi\, \underset{4}{c}_p(k) \tag{7. 76}$$

determines the equations of motion

$$\underset{4}{c}_p(k) = 0. \tag{7. 77}$$

But, from (7. 75), we have

$$-\underset{4}{\theta}_{pq} = \partial_r\, F_{pqr}, \qquad F_{pqr} = -F_{prq}, \tag{7. 78}$$

whence

$$\int \underset{4}{\theta}_{pk}\, n_k\, ds = -\int \partial_r\, F_{pkr}\, n_k\, ds \equiv 0, \tag{7. 79}$$

since this surface integral is over a rotational. Thus, the equations of motion vanish identically. It is therefore impossible to deduce the fourth order equations of motion of a charged particle in an electromagnetic field. According to Kursunoglu (107), this failure is due to the disappearance of the energy tensor. Einstein believes that it is still possible to represent matter by nonsingular solutions and that the problem of determining the Lorentz equations can still be stated. These equations will then represent a statistical phenomena involving the ensemble of the interactions between the particles: a rigorous solution which would account for this interaction will then explain as a global effect the motion of particles in an electromagnetic field.*

---

*Note added in proof. We note here a recent paper by Bonnor (Proc. Roy. Soc. **226A**, pp. 366-377, 1954). Bonnor succeeds in deducing the equations of motion from the equations of the unified field theory. The coulomb force is obtained at the price of complicating the Hamiltonian which becomes

$$\mathfrak{H}^* = \mathfrak{H} + p^2 \sqrt{-g}\, f^{ik} \varphi_{ik},$$

p being a real or imaginary constant. In certain aspects, this theory is similar to Kursunoglu's attempt (107) and leads to

$$m \frac{d\, r^2}{dt^2} = - \frac{m^{(1)}\, m^{(2)}}{r^3} \vec{r} + p^2 q^2 \frac{e^{(1)} e^{(2)}}{r^3} \vec{r}$$

After the usual quasi-static approximations have been made.

Appendix I

RELATIONS BETWEEN DETERMINANTS

Without difficulty, we can establish the relation

$$\gamma\gamma^{\mu\rho}\gamma^{\nu\sigma}\gamma^{\lambda\tau}\varphi_{\rho\lambda}\varphi_{\sigma\tau} = \frac{\gamma}{2}\gamma^{\mu\nu}\gamma^{\rho\sigma}\gamma^{\lambda\tau}\varphi_{\rho\lambda}\varphi_{\sigma\tau} - \varphi\varphi^{\mu\rho}\varphi^{\nu\sigma}\gamma_{\rho\sigma} \tag{I. 1}$$

and its reciprocal

$$\frac{1}{h}h_{\mu\rho}h_{\nu\sigma}h_{\lambda\tau}f^{\rho\lambda}f^{\sigma\tau} = \frac{1}{2h}h_{\mu\nu}h_{\rho\sigma}h_{\lambda\tau}f^{\rho\lambda}f^{\sigma\tau} - \frac{1}{f}f_{\mu\rho}f_{\nu\sigma}h^{\rho\sigma}. \tag{I. 2}$$

Substituting (1. 14) in (I. 1) and (1. 16) in (I. 2), we have

$$\gamma\gamma^{\mu\rho}\gamma^{\nu\sigma}\gamma^{\lambda\tau}\varphi_{\rho\lambda}\varphi_{\sigma\tau} = (g - \gamma - \varphi)\gamma^{\mu\nu} - \varphi\varphi^{\mu\rho}\varphi^{\nu\sigma}\gamma_{\rho\sigma}, \tag{I. 3}$$

$$\frac{1}{h}h_{\mu\rho}h_{\nu\sigma}h_{\lambda\tau}f^{\rho\lambda}f^{\sigma\tau} = \left(\frac{1}{g} - \frac{1}{h} - \frac{1}{f}\right)h_{\mu\nu} - \frac{1}{f}f_{\mu\rho}f_{\nu\sigma}h^{\rho\sigma} \tag{I. 4}$$

and the two other complementary relations (forming $\gamma_{\mu\pi}\gamma_{\nu\epsilon} \times$ (I. 3) and $h^{\mu\pi}h^{\nu\epsilon} \times$ (I. 4))

$$\varphi\gamma_{\mu\rho}\gamma_{\nu\sigma}\gamma_{\lambda\tau}\varphi^{\rho\lambda}\varphi^{\sigma\tau} = (g - \gamma - \varphi)\ \gamma_{\mu\nu} - \gamma\varphi_{\mu\rho}\varphi_{\nu\sigma}\gamma^{\rho\sigma}, \qquad (I.\ 5)$$

$$\frac{1}{f}h^{\mu\rho}h^{\nu\sigma}h^{\lambda\tau}f_{\rho\lambda}f_{\sigma\tau} = \left(\frac{1}{g} - \frac{1}{h} - \frac{1}{f}\right)h^{\mu\nu} - \frac{1}{h}f^{\mu\rho}f^{\nu\sigma}h_{\rho\sigma} \qquad (I.\ 6)$$

Multiplication of (I. 3) by $\gamma_{\nu\epsilon}$ and (I. 4) by $h^{\nu\epsilon}$ leads to the following relations (these result also from $\gamma^{\nu\epsilon} \times$ (I. 5) and $h_{\nu\epsilon} \times$ (I. 6):

$$\gamma\gamma^{\mu\rho}\gamma^{\lambda\tau}\varphi_{\rho\lambda}\varphi_{\epsilon\tau} = (g - \gamma - \varphi)\ \delta^{\mu}_{\epsilon} - \varphi\varphi^{\mu\rho}\varphi^{\nu\sigma}\gamma_{\rho\sigma}\gamma_{\nu\epsilon}, \qquad (I.7)$$

$$\frac{1}{h}h_{\mu\rho}h_{\lambda\tau}f^{\rho\lambda}f^{\epsilon\tau} = \left(\frac{1}{g} - \frac{1}{h} - \frac{1}{f}\right)\delta^{\epsilon}_{\mu} - \frac{1}{f}f_{\mu\rho}f_{\nu\sigma}h^{\rho\sigma}h^{\nu\epsilon}. \qquad (I.\ 8)$$

Contraction of ϵ and μ and use of (1. 14) and (1.16) gives:

$$\gamma\gamma^{\mu\rho}\gamma^{\nu\sigma}\varphi_{\mu\nu}\varphi_{\rho\sigma} = \varphi\varphi^{\mu\rho}\varphi^{\nu\sigma}\gamma_{\mu\nu}\gamma_{\rho\sigma} = 2(g - \gamma - \varphi), \qquad (I.\ 9)$$

$$\frac{1}{h}h_{\mu\rho}h_{\nu\sigma}f^{\mu\nu}f^{\rho\sigma} = \frac{1}{f}f_{\mu\rho}f_{\nu\sigma}h^{\mu\nu}h^{\rho\sigma} = 2\left(\frac{1}{g} - \frac{1}{h} - \frac{1}{f}\right) \qquad (I.\ 10)$$

1. DERIVATION OF THE RECIPROCAL RELATIONS (1.20) AND (1.21).

a. Calculation of $f_{\mu\nu}$ and $\varphi_{\mu\nu}$ – Substitution of (1.18a) in (1. 9) and use of (1. 8) and (1. 12) yields:

$$f_{\mu\nu} = \frac{\sqrt{f\varphi}}{g}(\varphi_{\mu\nu} + \gamma_{\mu\rho}\gamma_{\nu\sigma}\varphi^{\rho\sigma}). \qquad (I.\ 11)$$

In a similar way using (1.19a), (1. 8), (1. 9) and (1. 13), we have:

$$\varphi^{\mu\nu} = \frac{g}{\sqrt{f\varphi}}(f^{\mu\nu} + h^{\mu\rho}h^{\nu\sigma}f_{\rho\sigma}). \qquad (I.\ 12)$$

b. Calculation of $h_{\mu\nu}$ and $\gamma^{\mu\nu}$. –(1.18s) and (1.19s) can be written:

$$h_{\mu\nu} = \frac{h}{g}\gamma_{\mu\nu} - \frac{h}{f} f_{\mu\rho} f_{\nu\sigma} h^{\rho\sigma}, \qquad (I.13)$$

$$\gamma^{\mu\nu} = \frac{g}{\gamma} h^{\mu\nu} - \frac{\varphi}{\gamma} \varphi^{\mu\rho}\varphi^{\nu\sigma}\gamma_{\rho\sigma}. \qquad (I.14)$$

In these relations, we replace $f_{\mu\rho}$ and $\varphi^{\mu\rho}$ by (I. 11) and (I . 12) and $h^{\rho\sigma}$ and $\gamma_{\rho\sigma}$ by (1. 18s) and (1. 19s). We have after some computations and using (I. 5) through (I. 10):

$$h_{\mu\nu} = \frac{h\gamma}{g^2}(\gamma_{\mu\nu} + \varphi_{\mu\rho}\varphi_{\nu\sigma}\gamma^{\rho\sigma}), \qquad (I.15)$$

$$\gamma^{\mu\nu} = \frac{g^2}{h\gamma}(h^{\mu\nu} + f^{\mu\rho} f^{\nu\sigma} h_{\rho\sigma}). \qquad (I.16)$$

Multiplying (1. 18s) by (I.15) and (1. 18a) by (I. 11) (or (1. 19s) × (I. 16) and (1. 19a) × (I. 12)) leads to

$$g^2 = \gamma h = f\varphi \qquad (I.17)$$

We note that these conditions are such that

$$f(g^2 - \gamma h) = h(g^2 - f\varphi), \qquad (I.18)$$

which was obtained by multiplying (1. 18s) by (1. 19s), is automatically satisfied.

Substitution of (I. 17) into (I. 11), (I. 12), (I. 15) and (I. 16) leads to (1.20) and (1.21)

REMARK: Case $\varphi = 0$, $\frac{1}{f} = 0$. - (1. 8) and (1. 9) can then be written

$$2\sqrt{\varphi}\,\varphi^{\mu\nu} = \epsilon^{\mu\nu\rho\sigma}\varphi_{\rho\sigma}, \qquad \frac{2}{\sqrt{f}} f_{\mu\nu} = \epsilon_{\mu\nu\rho\sigma} f^{\rho\sigma} \qquad (I.19)$$

and, in this form, are always valid. We will then have from (1. 18a) and (1. 19a) with use of (1. 13)

$$2\sqrt{\varphi}\ \varphi^{\mu\nu} = \epsilon^{\mu\nu\rho\sigma}\varphi_{\rho\sigma} = \epsilon^{\mu\nu\rho\sigma}\ \frac{g}{h}\left(h_{\rho\lambda}h_{\sigma\tau}f^{\lambda\tau}\right)$$

$$= gh^{\mu\rho}h^{\nu\sigma}\epsilon_{\rho\sigma\lambda\tau}f^{\lambda\tau} = \frac{2g}{\sqrt{f}}\,h^{\mu\rho}h^{\nu\sigma}f_{\rho\sigma}. \qquad \text{(I. 20)}$$

whence

$$\sqrt{\varphi}\ \varphi^{\mu\nu} = \frac{g}{\sqrt{f}}\,h^{\mu\rho}h^{\nu\sigma}f_{\rho\sigma}. \qquad \text{(I. 21)}$$

Similarly, we have

$$\frac{1}{\sqrt{f}}\,f_{\mu\nu} = \frac{\sqrt{\varphi}}{g}\,\gamma_{\mu\rho}\gamma_{\nu\sigma}\,\varphi^{\rho\sigma}. \qquad \text{(I. 22)}$$

Appendix II

APPLICATION OF THE VARIATIONAL PRINCIPLE TO A DENSITY CONSTRUCTED WITH RICCI'S TENSOR($R_{\mu\nu}(L)$ (55) ($L_\rho = \underset{V}{L^{\sigma}_{\rho\sigma}} = 0$).)

We have assumed that the hamiltonian $\mathcal{G}^{\mu\nu}R_{\mu\nu}$ could be expressed as function of the Ricci tensor defined in terms of a general connection $\Gamma^{\rho}_{\mu\nu}$. Let us define the following affine connections

$$ {}^{(1)}L^{\rho}_{\mu\nu} = L^{\rho}_{\mu\nu} = \Gamma^{\rho}_{\mu\nu} + \frac{2}{3}\,\delta^{\rho}_{\mu}\,\Gamma_{\nu}\,, \qquad {}^{(1)}L_{\mu} = \underset{V}{{}^{(1)}L^{\rho}_{\mu\rho}} = 0, \tag{1.77} $$

$$ {}^{(2)}L^{\rho}_{\mu\nu} = \Gamma^{\rho}_{\mu\nu} + \frac{1}{3}(\delta^{\rho}_{\mu}\,\Gamma_{\nu} - \delta^{\rho}_{\nu}\Gamma_{\mu})\,, \qquad {}^{(2)}L_{\mu} = \underset{V}{{}^{(2)}L^{\rho}_{\mu\rho}} = 0, \tag{1.78} $$

and the three tensors

$$ {}^{(1)}R_{\mu\nu} = R_{\mu\nu}({}^{(1)}L) = \partial_{\rho}\,{}^{(1)}L^{\rho}_{\mu\nu} - \partial_{\nu}\,{}^{(1)}L^{\rho}_{\mu\rho} + {}^{(1)}L^{\lambda}_{\mu\nu}\,{}^{(1)}L^{\rho}_{\lambda\rho} - {}^{(1)}L^{\lambda}_{\mu\rho}\,{}^{(1)}L^{\rho}_{\lambda\nu}, \tag{II. 1$_1$} $$

$$ {}^{(2)}R_{\mu\nu} = R_{\mu\nu}({}^{(2)}L) = \partial_\rho {}^{(2)}L^{\rho}_{\mu\nu} - \partial_\nu {}^{(2)}L^{\rho}_{\mu\rho} + {}^{(2)}L^{\lambda}_{\mu\nu} {}^{(2)}L^{\rho}_{\lambda\rho} - {}^{(2)}L^{\lambda}_{\mu\rho} {}^{(2)}L^{\rho}_{\lambda\nu}, \qquad (\text{II. } 1_2) $$

$$ {}^{(3)}R_{\mu\nu} = \frac{1}{2}\left[{}^{(2)}R_{\mu\nu} + {}^{(2)}\tilde{R}_{\nu\mu}\right]. \qquad (\text{II. } 1_3) $$

In terms of the initial connection $\Gamma^{\rho}_{\mu\nu}$, these three tensors have the form

$$ {}^{(1)}R_{\mu\nu} = W_{\mu\nu} = R_{\mu\nu} + \frac{2}{3}(\partial_\mu \Gamma_\nu - \partial_\nu \Gamma_\mu), \qquad (\text{II. } 2_1) $$

$$ {}^{(2)}R_{\mu\nu} = \partial_\rho \Gamma^{\rho}_{\mu\nu} - \partial_\nu \Gamma^{\rho}_{\underline{\mu\rho}} + \Gamma^{\lambda}_{\mu\nu}\Gamma^{\rho}_{\underline{\lambda\rho}} - \Gamma^{\lambda}_{\mu\rho}\Gamma^{\rho}_{\lambda\nu} + \frac{1}{3}(\partial_\mu \Gamma_\nu - \partial_\nu \Gamma_\mu) - \frac{1}{3}\Gamma_\mu \Gamma_\nu, \qquad (\text{II. } 2_2) $$

$$ {}^{(3)}R_{\mu\nu} = {}^{(2)}R_{\mu\nu} - \frac{1}{2}(\partial_\mu \Gamma^{\rho}_{\underline{\nu\rho}} - \partial_\nu \Gamma^{\rho}_{\underline{\mu\rho}}). \qquad (\text{II. } 2_3) $$

With the aid of these tensors, we can construct the three hamiltonians $\mathcal{G}^{\mu\nu}\,{}^{(a)}R_{\mu\nu}$ ($a = 1, 2$ or 3). If the ${}^{(a)}R_{\mu\nu}$ are expressed as functions of $L^{\rho}_{\mu\nu}$ as in (II. 1a)., we must use Lagrange multipliers to take into account the conditions (1. 77) and (1. 78). We will obtain the same results by expressing ${}^{(a)}R_{\mu\nu}$ in terms of $\Gamma^{\rho}_{\mu\nu}$ (II. 2a) and applying to these expressions the variational principle developed in Chapter 2 (B, § 3). Having obtained the field equations, we can transform back to the connections ${}^{(1)}L$ or ${}^{(2)}L$ which are the connections with vanishing torsions occurring in the hamiltonians. We thus obtain either

$$ {}^{(1)}D_\rho \mathcal{G}^{\mu\nu}_{+-} = -\frac{2}{3}\delta^{\nu}_{\rho}\, \mathfrak{g}^{\mu}, \qquad (\text{II. } 3_1) $$

$$ {}^{(2)}D_\rho \mathcal{G}^{\mu\nu}_{+-} = -\frac{2}{3}\delta^{\mu}_{\rho}\, \mathfrak{g}^{\nu}, \qquad (\text{II. } 3_2) $$

or

$$\overset{(2)}{D}_\rho \mathcal{G}^{\underset{+-}{\mu\nu}} = \frac{1}{3}(\delta^\nu_\rho \mathcal{J}^\mu - \delta^\mu_\rho F^\nu) \qquad \text{(II. 3}_3\text{)}$$

and in general

$$\overset{(a)}{R}_{\mu\nu} = 0 \qquad \text{(II. 3)}$$

In the above $\overset{(1)}{D}_\rho$ and $\overset{(2)}{D}_\rho$ represent the covariant derivatives with respect to $\overset{(1)}{L}{}^\rho_{\mu\nu}$ and $\overset{(2)}{L}{}^\rho_{\mu\nu}$. Using the tensors instead of the densities, we have

$$\overset{(1)}{D}_\rho g^{\underset{+-}{\mu\nu}} = -\frac{2}{3}\delta^\nu_\rho f^\mu + \frac{1}{3} g^{\mu\nu} g_{\sigma\rho} f^\sigma, \qquad \text{(II. 4}_1\text{)}$$

$$\overset{(2)}{D}_\rho g^{\underset{+-}{\mu\nu}} = -\frac{2}{3}\delta^\mu_\rho f^\nu + \frac{1}{3} g^{\mu\nu} g_{\rho\sigma} f^\sigma, \qquad \text{(II. 4}_2\text{)}$$

$$\overset{(2)}{D}_\rho g^{\underset{+-}{\mu\nu}} = -\frac{1}{3}(\delta^\nu_\rho f^\mu - \delta^\mu_\rho f^\nu) - \frac{1}{3} g^{\mu\nu} \varphi_{\rho\sigma} f^\sigma. \qquad \text{(II. 4}_3\text{)}$$

It is now a simple matter to define a new affine connection Δ such that the covariant derivative of $g^{\mu\nu}$ vanishes. Depending on whether we choose $\overset{(1)}{R}_{\mu\nu}$, $\overset{(2)}{R}_{\mu\nu}$ or $\overset{(3)}{R}_{\mu\nu}$, we have

$$\overset{(1)}{L}{}^\rho_{\mu\nu} = \overset{(1)}{\Delta}{}^\rho_{\mu\nu} - \frac{1}{2} g_{\mu\nu}(f^\rho - \bar{f}^\rho) - \frac{1}{6}\delta^\rho_\mu g_{\lambda\nu} f^\lambda \qquad \text{(II.5}_1\text{)}$$
$$- \frac{1}{2}\delta^\rho_\mu g_{\lambda\nu} \bar{f}^\lambda + \frac{1}{2}\delta^\rho_\nu g_{\lambda\mu}(f^\lambda + \bar{f}^\lambda),$$

$$\overset{(2)}{L}{}^\rho_{\mu\nu} = \overset{(2)}{\Delta}{}^\rho_{\mu\nu} - \frac{1}{2} g_{\mu\nu}(f^\rho + \bar{f}^\rho) - \frac{1}{6}\delta^\rho_\nu g_{\mu\lambda} f^\lambda \qquad \text{(II. 5}_2\text{)}$$
$$+ \frac{1}{2}\delta^\rho_\nu g_{\mu\nu} \bar{f}^\lambda + \frac{1}{2}\delta^\rho_\mu g_{\nu\lambda}(f^\lambda - \bar{f}^\lambda),$$

$$\overset{(2)}{L}{}^\rho_{\mu\nu} = \overset{(3)}{\Delta}{}^\rho_{\mu\nu} - \frac{1}{2} g_{\mu\nu} \bar{f}^\rho - \frac{1}{6}(\delta^\rho_\mu g_{\lambda\nu} - \delta^\rho_\nu g_{\mu\lambda}) f^\lambda \qquad \text{(II.5}_3\text{)}$$
$$+ \frac{1}{2}(\delta^\rho_\mu g_{\nu\lambda} - \delta^\rho_\nu g_{\mu\lambda})(f^\lambda - \bar{f}^\lambda),$$

with

$$f^{\rho} = \frac{1}{\sqrt{-g}}\, \partial_{\mu}\, \mathfrak{g}^{\rho\mu}, \quad f^{\bar{\rho}} = \gamma^{\rho\sigma}\varphi_{\sigma\tau} f^{\tau}. \tag{II. 6}$$

Let $;\rho$ designate a covariant derivative with respect to Δ. Then the field equations (II. 3a) and (II. 3) can be written *

$$\mathfrak{G}^{\underset{+-}{\mu\rho}}{}_{;\rho} = 0, \quad {}^{(1)}R_{\mu\nu} = R_{\mu\nu}({}^{(1)}\Delta) + {}^{(1)}K_{\mu\nu} = 0, \tag{II. 7$_1$}$$

$$\mathfrak{G}^{\underset{+-}{\mu\nu}}{}_{;\rho} = 0, \quad {}^{(2)}R_{\mu\nu} = R_{\mu\nu}({}^{(2)}\Delta) + {}^{(2)}K_{\mu\nu} = 0, \tag{II. 7$_2$}$$

$$\mathfrak{G}^{\underset{+-}{\mu\nu}}{}_{;\rho} = 0, \quad {}^{(3)}R_{\mu\nu} = R_{\mu\nu}({}^{(3)}\Delta) + {}^{(3)}K_{\mu\nu} = 0, \tag{II. 7$_3$}$$

with

$$R_{\mu\nu}(\Delta) = \partial_{\rho}\Delta^{\rho}_{\mu\nu} - \partial_{\nu}\Delta^{\rho}_{\underline{\mu\rho}} + \Delta^{\lambda}_{\mu\nu}\Delta^{\rho}_{\underline{\lambda\rho}} - \Delta^{\lambda}_{\mu\rho}\,\Delta^{\rho}_{\lambda\nu}\,. \tag{II. 8}$$

and

$$\begin{aligned} {}^{(1)}K_{\mu\nu} = {}^{(2)}K_{\mu\nu} = \frac{-1}{2\sqrt{-g}}\, g_{\mu\nu}\partial_{\rho}\, \mathfrak{g}^{\bar{\rho}} & \\ + \frac{1}{2}[\partial_{\mu}(f_{\nu} - f_{\bar{\bar{\nu}}}) - \partial_{\nu}(f_{\mu} - f_{\bar{\bar{\mu}}})] & \\ - \frac{1}{2}(f_{\mu} - f_{\bar{\bar{\mu}}})(f_{\nu} - f_{\bar{\bar{\nu}}}), & \end{aligned} \tag{II. 9}$$

$$\begin{aligned} {}^{(3)}K_{\mu\nu} = {}^{(1)}K_{\mu\nu} - \frac{1}{6}(\partial_{\mu}f_{\nu} - \partial_{\nu}f_{\mu}) & \\ + \frac{1}{6}(f_{\mu} + f_{\bar{\mu}})(f_{\nu} - f_{\bar{\nu}}). & \end{aligned} \tag{II. 10}$$

---

* To verify these relations, we use

$$g^{\sigma\nu} g_{\lambda\sigma}(f^{\lambda} + f^{\bar{\lambda}}) = f^{\nu} - f^{\bar{\nu}}, \quad g^{\nu\sigma} g_{\sigma\lambda}(f^{\lambda} - f^{\bar{\lambda}}) = f^{\nu} + f^{\bar{\nu}}$$

or

$$g_{\mu\nu}(f^{\nu} - f^{\bar{\nu}}) = g_{\nu\mu}(f^{\nu} + f^{\bar{\nu}}).$$

The connections $\overset{(a)}{\Delta}{}^{\rho}_{\mu\nu}$ satisfy the relation $\mathcal{G}^{\underset{+-}{\mu\nu}}{}_{;\rho} = 0$. Thus we must have† (cf. (2.41) from (1.51)

$$\Delta_{\mu} = \overset{(a)}{\underset{V}{\Delta_{\mu\rho}}} = g_{\mu\rho}(f^{\rho} - f^{\bar{\rho}}) = f_{\mu} - f_{\bar{\bar{\mu}}}, \qquad \text{(II. 11)}$$

with

$$f_{\mu} = \gamma_{\mu\rho} f^{\rho}, \qquad f_{\bar{\bar{\mu}}} = \varphi_{\mu\sigma}\gamma^{\sigma\lambda}\varphi_{\lambda\nu} f^{\nu}. \qquad \text{(II. 12)}$$

We can show that (II. 11) follows form (II. 5a). If we want to impose the a priori conditions $\partial_{\rho}\mathcal{G}^{\mu\rho} = 0$, we must introduce into the variational principle Lagrange's multipliers. In this case, Δ becomes the same as L and the field equations are identical to Eq. II of Chapter 2.

---

† $\partial_{\rho}\mathcal{G}^{\mu\rho} - \mathcal{K}^{\mu\rho}\Delta_{\rho} = 0$ has as a consequence

$$\Delta_{\mu} = h_{\mu\nu} f^{\nu} = (\gamma_{\mu\nu} + \varphi_{\mu\sigma}\varphi_{\nu\sigma}\gamma^{\rho\sigma}) f^{\nu} = f_{\mu} - f_{\bar{\bar{\mu}}}$$

We also have

$$g_{\mu\nu}(f^{\nu} - f^{\bar{\nu}}) = (\gamma_{\mu\nu} + \varphi_{\mu\nu})(f^{\nu} - \gamma^{\nu\sigma}\varphi_{\sigma\tau} f^{\tau}) = (\gamma_{\mu\nu} + \varphi_{\mu\sigma}\varphi_{\nu\tau}\gamma^{\sigma\tau}) f^{\nu}$$

$$= h_{\mu\nu} f^{\nu} = f_{\mu} - f_{\bar{\bar{\mu}}},$$

Similarily

$$g_{\nu\mu}(f^{\nu} + f^{\bar{\nu}}) = h_{\nu\mu} f^{\nu} = f_{\mu} - f_{\bar{\bar{\mu}}},$$

with

$$f_{\bar{\bar{\mu}}} = \varphi_{\mu\sigma}\gamma^{\sigma\tau}\varphi_{\tau\nu} f^{\nu}.$$

Appendix III

PROOF OF THE RELATION

$$\left(A_{\overline{\overline{\overline{\rho}}}} = - \frac{\varphi}{\gamma} A_{\rho} - \frac{1}{\gamma}(g - \gamma - \varphi) A_{\overline{\overline{\rho}}}. \right)$$

We start from

$$A_{\overline{\overline{\overline{\rho}}}} = \varphi_{\rho\sigma} \gamma^{\sigma\tau} \varphi_{\tau\lambda} \gamma^{\lambda\mu} \varphi_{\mu\pi} \gamma^{\pi\delta} A_{\delta} \qquad \text{(III. 1)}$$

Interchanging τ and λ, and σ and μ, we have:

$$A_{\overline{\overline{\overline{\rho}}}} = \frac{1}{2}(\varphi_{\rho\sigma}\varphi_{\mu\pi} - \varphi_{\rho\mu}\varphi_{\sigma\pi} - \varphi_{\rho\pi}\varphi_{\mu\sigma}) \gamma^{\sigma\tau}\gamma^{\lambda\mu} \varphi_{\tau\lambda}\gamma^{\pi\delta} A_{\delta}$$
$$+ \frac{1}{2} \varphi_{\rho\pi}\varphi_{\mu\sigma}\gamma^{\sigma\tau}\gamma^{\lambda\mu} \varphi_{\tau\lambda}\gamma^{\pi\delta} A_{\delta}. \qquad \text{(III. 2)}$$

But from the value of φ (cf. (1. 10)) and from (1. 14):

$$\varphi_{\rho\sigma}\varphi_{\mu\pi} - \varphi_{\rho\mu}\varphi_{\sigma\pi} - \varphi_{\rho\pi}\varphi_{\mu\sigma} = \sqrt{\varphi}\ \epsilon_{\rho\sigma\mu\pi}, \qquad \text{(III. 3)}$$

$$\frac{1}{2} \gamma^{\sigma\tau}\gamma^{\lambda\mu} \varphi_{\mu\sigma}\varphi_{\tau\lambda} = - \frac{1}{\gamma}(g - \gamma - \varphi). \qquad \text{(III. 4)}$$

Thus

$$A_{\overline{\overline{\overline{\rho}}}} = \frac{\sqrt{\varphi}}{2} \epsilon_{\rho\sigma\mu\pi} \gamma^{\sigma\tau}\gamma^{\lambda\mu} \varphi_{\tau\lambda}\gamma^{\pi\delta} A_{\delta} - \frac{1}{\gamma}(g - \gamma - \varphi) A_{\bar{\rho}}, \qquad \text{(III. 5)}$$

that is from (1. 12), (1. 6) and (1. 8)

$$A_{\bar{\bar{\bar{\rho}}}} = \frac{\sqrt{\varphi}}{2\gamma}\epsilon^{\mu\pi\lambda\delta}\gamma_{\rho\mu}\gamma_{\sigma\pi}\gamma^{\sigma\tau}\varphi_{\tau\lambda}A_{\delta} - \frac{1}{\gamma}(g - \gamma - \varphi)\,A_{\bar{\rho}}$$

$$= \frac{\sqrt{\varphi}}{2\gamma}\epsilon^{\mu\tau\lambda\delta}\gamma_{\rho\mu}\varphi_{\tau\lambda}A_{\delta} - \frac{1}{\gamma}(g - \gamma - \varphi)\,A_{\bar{\rho}} \qquad \text{(III. 6)}$$

$$= \frac{\varphi}{\gamma}\varphi^{\mu\delta}\gamma_{\rho\mu}\,A_{\delta} - \frac{1}{\gamma}(g - \gamma - \varphi)\,A_{\bar{\rho}}.$$

Now $A_{\bar{\bar{\bar{\bar{\rho}}}}}$ is

$$A_{\bar{\bar{\bar{\bar{\rho}}}}} = \varphi_{\rho\sigma}\gamma^{\sigma\tau}A_{\bar{\bar{\bar{\tau}}}}. \qquad \text{(III. 7)}$$

But from (III.6)

$$A_{\bar{\bar{\bar{\rho}}}} = \frac{\varphi}{\gamma}\,\varphi_{\rho\sigma}\gamma^{\sigma\tau}\gamma_{\tau\mu}\varphi^{\mu\delta}A_{\delta} - \frac{1}{\gamma}(g - \gamma - \varphi)\,\varphi_{\rho\sigma}\gamma^{\sigma\tau}A_{\bar{\rho}} \qquad \text{(III 8)}$$

or by using (1. 6)

$$A_{\bar{\bar{\bar{\bar{\rho}}}}} = -\frac{\varphi}{\gamma}A_{\rho} - \frac{1}{\gamma}(g - \gamma - \varphi)\,A_{\bar{\bar{\rho}}}. \qquad \text{(III. 9)}$$

Appendix IV

CALCULATION OF THE AFFINE CONNECTION IN THE STATIC SPHERICALLY SYMMETRIC CASE (78)

1. COMPUTATION OF $R_{\underset{\vee}{\mu\nu},\rho}$.

The $R_{\mu\nu,\rho}$ are given by the formulas (3. 56). In the case under consideration, $g^{\mu\nu}$ and $g_{\mu\nu}$ have the values given in (5. 14) and (5. 27). Thus the only non-vanishing component of $\varphi_{\underline{\mu\nu\lambda}}$ is φ_{123} and the only non-vanishing components of $\varphi^*_{[\mu\nu]\epsilon} = \frac{\sqrt{-\gamma}}{2}\,\epsilon_{\mu\nu\rho\sigma}\gamma^{\rho\lambda}\gamma^{\sigma\tau}\varphi_{\lambda\tau\epsilon}$ will be $\varphi^*_{[14]\,1}$, $\varphi^*_{[24]\,2}$, $\varphi^*_{[34]\,3}$. On the other hand only f_4 and Δ_4 are different from zero. From (3. 56), the only non-vanishing components of $R_{\underset{\vee}{\mu\nu},\rho}$ in the static case are:

$$\underset{V}{R_{23,1}} = -\frac{1}{2}\varphi_{123} + \nabla_1\varphi_{23} + \frac{\sqrt{\varphi}}{2\sqrt{-\gamma}}\varphi^*_{23}\left(\varphi^{23}\varphi_{123} + \partial_1 \operatorname{Log}\frac{g}{\varphi}\right)$$

$$+\left(\frac{\sqrt{\varphi}}{2}\varphi^{14} - \varphi_{23}\right)\partial_1 \operatorname{Log}\frac{g}{\gamma} + \gamma^{44}(\sqrt{\varphi} + \varphi_{41}\varphi_{23})\Delta_4,$$

$$\underset{V}{R_{14,1}} = \nabla_1\varphi_{14} + \frac{\sqrt{\varphi}}{2\sqrt{-\gamma}}\varphi^*_{[14]\ 1} + \frac{\sqrt{\varphi}}{2\sqrt{-\gamma}}\varphi^*_{14}\left(\varphi^{23}\varphi_{123} + \partial_1 \operatorname{Log}\frac{g}{\varphi}\right)$$

$$-\varphi_{14}\partial_1 \operatorname{Log}\frac{g}{\varphi} - \frac{\varphi}{2}\gamma^{22}\gamma^{33}\varphi^{14}\partial_1 \operatorname{Log}\frac{g}{\varphi},$$

$$\underset{V}{R_{31,2}} = -\frac{1}{2}\varphi_{123} - \begin{Bmatrix}2\\12\end{Bmatrix}\varphi_{32} + \frac{\sqrt{\varphi}}{2}\varphi^{14}\partial_1 \operatorname{Log}\frac{g}{\gamma} + \sqrt{\varphi}\,\gamma^{44}\,\Gamma_4,$$

$$\underset{V}{R_{24,2}} = -\begin{Bmatrix}1\\22\end{Bmatrix}\varphi_{14} + \frac{\sqrt{\varphi}}{2\sqrt{-\gamma}}\varphi^*_{[24]\ 2}$$

$$-\frac{\varphi}{2\sqrt{-\gamma}}\epsilon^*_{[24]\ 24}\varphi^{14}\partial_1 \operatorname{Log}\frac{g}{\varphi} + \gamma^{33}\varphi_{32}\varphi_{32}\Delta_4,$$

$$\underset{V}{R_{12,3}} = -\frac{1}{2}\varphi_{123} - \begin{Bmatrix}3\\13\end{Bmatrix}\varphi_{32} + \frac{\sqrt{\varphi}}{2}\varphi^{14}\partial_1 \operatorname{Log}\frac{g}{\gamma} + \sqrt{\varphi}\,\gamma^{44}\Delta_4,$$

$$\underset{V}{R_{34,3}} = -\begin{Bmatrix}1\\33\end{Bmatrix}\varphi_{14} + \frac{\sqrt{\varphi}}{2\sqrt{-\gamma}}\varphi^*_{[34]\ 3}$$

$$-\frac{\varphi}{2\sqrt{-\gamma}}\epsilon^*_{[34]\ 34}\varphi^{14}\partial_1 \operatorname{Log}\frac{g}{\gamma} + \gamma^{22}\varphi_{23}\varphi_{23}\Delta_4. \qquad \text{(IV. 1)}$$

From the values of $\gamma_{\mu\nu}$ and $\varphi_{\mu\nu}$, we see that

$$\underset{V}{R_{12,3}} = \underset{V}{R_{31,2}}, \qquad \underset{V}{R_{34,3}} = \underset{V}{R_{24,2}}\sin^2\theta. \qquad \text{(IV. 2)}$$

But

$$f^4 = \frac{1}{\sqrt{-g}}\partial_1\left(\sqrt{-g}\,f^{41}\right)$$

$$= \frac{1}{(\alpha\sigma - w^2)^{\frac{1}{2}}(\beta^2 + u^2)}\partial_1\frac{w(\beta^2 + u^2)^{\frac{1}{2}}}{(\alpha\sigma - w^2)^{\frac{1}{2}}}, \qquad \text{(IV. 3)}$$

since

$$f^{41} = \frac{\varphi}{g}\varphi^{41} + \frac{\gamma}{g}\gamma^{44}\gamma^{11}\varphi_{41} = \frac{w}{\alpha\sigma - w^2}, \qquad (IV.4)$$

we will have

$$f_4 = \gamma_{44} f^4, \quad f_{\bar{\bar{4}}} = \varphi_{41}\gamma^{11}\varphi_{14}\gamma^{44} f_4 = \frac{w^2}{\alpha\sigma} f_4,$$

whence

$$\Delta_4 = f_4 - f_{\bar{\bar{4}}} = \frac{\sigma(\alpha\sigma - w^2)}{\alpha\sigma(\alpha\sigma - w^2)^{\frac{1}{2}}(\beta^2 + u^2)^{\frac{1}{2}}}\,\partial_1 \frac{w(\beta^2 + u^2)^{\frac{1}{2}}}{(\alpha\sigma - w^2)^{\frac{1}{2}}}$$

$$= \frac{w}{\alpha}\,\partial_1 \operatorname{Log} \frac{w(\beta^2 + u^2)^{\frac{1}{2}}}{(\alpha\sigma - w^2)^{\frac{1}{2}}}. \qquad (IV.6)$$

Substitution of the values of $\gamma_{\mu\nu}$, $\varphi_{\mu\nu}$, $\varphi_{\nu\mu\rho}$ and Δ_4 into (IV. 1) leads to

$$\underset{V}{R_{23,1}} = -\frac{u}{2}\left(1 + \frac{w^2}{\alpha\sigma}\right)\partial_1 \operatorname{Log}\frac{(\alpha\sigma - w^2)(\beta^2 + u^2)}{(\alpha\sigma + w^2)u}\sin\theta,$$

$$\underset{V}{R_{14,1}} = \frac{w}{2}\left[\partial_1 \operatorname{Log}\frac{w^2}{\alpha\sigma} + \frac{u^2}{\beta^2}\partial_1 \operatorname{Log} u^2 - 2\partial_1 \operatorname{Log}\frac{(\alpha\sigma - w^2)(\beta^2 + u_2)}{\alpha\sigma\beta^2}\right],$$

$$\underset{V}{R_{31,2}} = \underset{V}{R_{12,3}} = \frac{u}{2}\left[\partial_1 \operatorname{Log}\frac{\beta}{u}\frac{\beta^2 + u^2}{\beta^2} + \frac{w^2}{\alpha\sigma}\partial_1 \operatorname{Log}(\beta^2 + u^2)\right]\sin\theta,$$

$$\underset{V}{R_{24,2}} = \frac{\underset{V}{R_{34,3}}}{\sin^2\theta} = \frac{w}{2}\frac{u^2}{\alpha\beta}\left[\left(1 + \frac{\beta^2}{u^2}\right)\partial_1 \operatorname{Log}\beta - \partial_1 \operatorname{Log}\frac{\beta^2 + u^2}{u^2}\frac{\beta}{u}\right]. \qquad (IV.7)$$

2. CALCULATION OF $\underset{V}{\Delta_{\mu\nu,\rho}}$ AS A FUNCTION OF $\underset{V}{R_{\mu\nu,\rho}}$.

From (3. 53), (3. 54) and (IV.7), we see that only six of the 24 antisymmetric coefficients of the affine connection do not vanish. From these six values of $\underset{V}{R_{\mu\nu,\rho}}$, we can easily construct the quantities $\underset{V}{R_{\mu\nu,\bar{\bar{\rho}}}}$ and $\underset{V}{R^*_{\mu\nu,\rho}}$ taking into account the values (5. 23 and (5. 27) of $\gamma^{\mu\nu}$ and $\varphi_{\mu\nu}$. As an example, we have

$$\left.\begin{aligned} \underset{V}{R}_{23,\bar{\bar{1}}} &= \varphi_{14}\gamma^{44}\varphi_{41}\gamma^{11}\underset{V}{R}_{23,1} = \frac{w^2}{\alpha\sigma}\underset{V}{R}_{23,1}, \\ \underset{V}{R}^{*}_{23,1} &= \sqrt{-\gamma}\,\epsilon_{2314}\gamma^{11}\gamma^{44}\underset{V}{R}_{14,1} = \frac{-\beta}{\sqrt{\alpha\sigma}}\underset{V}{R}_{14,1}\sin\theta, \\ &\dots\dots\dots\dots\dots\dots\dots\dots, \end{aligned}\right\} \quad (\text{IV. } 8)$$

We can thus calculate the six quantities $\underset{V}{S}_{\mu\nu,\rho}$ defined by (3.53). Using (IV. 8) and

$$2 - \frac{g}{\gamma} + \frac{\varphi}{\gamma} = \frac{\alpha\sigma\beta^2 - \alpha\sigma u^2 + w^2\beta^2}{\alpha\sigma\beta^2}, \quad \frac{2\sqrt{\varphi}}{\sqrt{-\gamma}} = \frac{2wu}{\beta\sqrt{\alpha\sigma}}. \quad (\text{IV. } 9)$$

we thus have:

$$\begin{aligned} \underset{V}{S}_{23,1} &= \left(1 - \frac{u^2}{\beta^2}\right)\underset{V}{R}_{23,1} + \frac{2wu}{\alpha\beta}\underset{V}{R}_{14,1}, \\ \underset{V}{S}_{14,1} &= \left(1 - \frac{u^2}{\beta^2}\right)\underset{V}{R}_{14,1} - \frac{2wu}{\beta^2\sin\theta}\underset{V}{R}_{23,1}, \qquad (\text{IV. } 10) \\ \underset{V}{S}_{12,3} = \underset{V}{S}_{31,2} &= \left(1 + \frac{w^2}{\alpha\sigma}\right)\underset{V}{R}_{31,2} + \frac{2wu}{\beta\sigma}\underset{V}{R}_{24,2}\sin\theta, \\ \underset{V}{S}_{24,2} = \frac{\underset{V}{S}_{34,3}}{\sin^2\theta} &= \left(1 + \frac{w^2}{\alpha\sigma}\right)\underset{V}{R}_{24,2} - \frac{2wu}{\alpha\beta\sin\theta}\underset{V}{R}_{31,2}. \end{aligned}$$

Substitution of $\underset{V}{S}_{\mu\nu,\rho}$ from (IV. 10) into (3. 54) and use of

$$\left.\begin{aligned} a = 2 - \frac{g}{\gamma} + \frac{\varphi}{\gamma} &= \left(1 + \frac{w^2}{\alpha\sigma}\right)\left(1 - \frac{u^2}{\beta^2}\right) - \frac{4w^2u^2}{\alpha\sigma\beta^2}, \\ b = \frac{2wu}{\beta\sqrt{\alpha\sigma}} &\left[\left(1 + \frac{w^2}{\alpha\sigma}\right) + \left(1 - \frac{u^2}{\beta^2}\right)\right], \end{aligned}\right\} \quad (\text{IV. } 11)$$

$$a^2 + b^2 = \left[\left(1 - \frac{u^2}{\beta^2}\right)^2 + \frac{4w^2u^2}{\alpha\sigma\beta^2}\right]\left[\left(1 + \frac{w^2}{\alpha\sigma}\right)^2 + \frac{4w^2u^2}{\alpha\sigma\beta^2}\right],$$

leads to

$$\left[\left(1 + \frac{w^2}{\alpha\sigma}\right)^2 + \frac{4w^2 u^2}{\alpha\sigma \beta^2}\right] \underset{v}{\Delta_{23,1}} = \left(1 + \frac{w^2}{\alpha\sigma}\right) \underset{v}{R_{23,1}} - \frac{2wu}{\alpha\sigma} \underset{v}{R_{14,1}} \sin\theta,$$

$$\left[\left(1 + \frac{w^2}{\alpha\sigma}\right)^2 + \frac{4w^2 u^2}{\alpha\sigma \beta^2}\right] \underset{v}{\Delta_{14,1}} = \left(1 + \frac{w^2}{\alpha\sigma}\right) \underset{v}{R_{14,1}} + \frac{2wu}{\beta^2 \sin\theta} \underset{v}{R_{23,1}},$$

$$\left[\left(1 - \frac{u^2}{\beta^2}\right)^2 + \frac{4w^2 u^2}{\alpha\sigma \beta^2}\right] \underset{v}{\Delta_{31,2}} = \left(1 - \frac{u^2}{\beta^2}\right) \underset{v}{R_{31,2}} - \frac{2wu}{\beta\sigma} \underset{v}{R_{24,2}} \sin\theta,$$

$$\left[\left(1 - \frac{u^2}{\beta^2}\right)^2 + \frac{4w^2 u^2}{\alpha\sigma\beta^2}\right] \underset{v}{\Delta_{24,2}} = \left(1 - \frac{u^2}{\beta^2}\right) \underset{v}{R_{24,2}} + \frac{2wu}{\alpha\beta \sin\theta} \underset{v}{R_{31,2}},$$

$$\underset{v}{\Delta_{12,3}} = \underset{v}{\Delta_{31,2}}, \qquad \underset{v}{\Delta_{34,3}} = \underset{v}{\Delta_{34,2}} \sin^2\theta.$$

$$(\text{IV. }12)$$

Substituting for $\underset{v}{R_{\mu\nu,\rho}}$ their values from (IV. 7), we get

$$\underset{v}{\Delta_{23,1}} = \frac{-u}{4}\left[\partial_1 \text{Log } \beta^2 \left(1 + \frac{u^2}{\beta^2}\right) - \frac{\beta^2}{u^2} \partial_1 \text{Log}\left(1 + \frac{u^2}{\beta^2}\right)\right] \sin\theta,$$

$$\underset{v}{\Delta_{14,1}} = \frac{-\alpha\sigma}{2w} \partial_1 \text{Log}\left(1 - \frac{w^2}{\alpha\sigma}\right),$$

$$(\text{IV. }13)$$

$$\underset{v}{\Delta_{31,2}} = \underset{v}{\Delta_{12,3}} = -\frac{\beta^2}{4u} \partial_1 \text{Log}\left(1 + \frac{u^2}{\beta^2}\right),$$

$$\underset{v}{\Delta_{24,2}} = \frac{\underset{v}{\Delta_{34,3}}}{\sin^2\theta} = \frac{w\beta}{4\alpha} \partial_1 \text{Log } \beta^2 \left(1 + \frac{u^2}{\beta^2}\right),$$

if the condition

$$a^2 + b^2 = \left[\left(1 - \frac{u^2}{\beta^2}\right)^2 + \frac{4w^2 u^2}{\alpha\sigma\beta^2}\right]$$

$$\times \left[\left(1 + \frac{w^2}{\alpha\sigma}\right)^2 + \frac{4w^2 u^2}{\alpha\sigma\beta^2}\right] \neq 0$$

$$(\text{IV. }14)$$

is satisfied. The $\underset{\vee}{\Delta}_{\mu\nu,\rho}$ is given by (IV. 13) are uniquely determined if

$$g = \left(1 + \frac{u^2}{\beta^2}\right)\left(1 - \frac{w^2}{\alpha\sigma}\right) \neq 0 \qquad \text{(IV. 15)}$$

The condition for existence then is

$$g(a^2 + b^2) = \left(1 + \frac{u^2}{\beta^2}\right)\left(1 - \frac{w^2}{\alpha}\right) \times \left[\left(1 - \frac{u^2}{\beta^2}\right)^2 + \frac{4w^2u^2}{\alpha\sigma\beta^2}\right]\left[\left(1 + \frac{w^2}{\alpha\sigma}\right)^2 + \frac{4w^2u^2}{\alpha\sigma\beta^2} \neq 0\right] \qquad \text{(IV. 16)}$$

We can give to (IV. 13) a slightly different form by introducing Bonnor's expressions(63):

$$A_1 = \frac{1}{2}\partial_1 \operatorname{Log} \beta^2 \left(1 + \frac{u^2}{\beta^2}\right) = \frac{uu' + \beta\beta'}{u^2 + \beta^2},$$
$$B_1 = -\frac{\beta}{2u}\partial_1 \operatorname{Log}\left(1 + \frac{u^2}{\beta^2}\right) = \frac{u\beta' - \beta u'}{u^2 + \beta^2}. \qquad \text{(IV. 17)}$$

Raising the last subscript in (IV. 13), we have

$$\underset{\vee}{\Delta}{}^{1}_{23} = \frac{\beta B_1 + uA_1}{2\alpha}, \qquad \underset{\vee}{\Delta}{}^{1}_{14} = \frac{\sigma}{2w}\partial_1 \operatorname{Log}\left(1 - \frac{w^2}{\alpha\sigma}\right), \qquad \text{(IV. 18)}$$
$$\underset{\vee}{\Delta}{}^{2}_{31} = \underset{\vee}{\Delta}{}^{3}_{12}\sin^2\theta = -\frac{B_1}{2}\sin\theta, \qquad \underset{\vee}{\Delta}{}^{2}_{24} = \underset{\vee}{\Delta}{}^{3}_{34} = -\frac{w}{2\alpha}A_1.$$

Let us now compute the symmetric part of the affine connection

$$\underline{\Delta}{}^{\rho}_{\mu\nu} = \left\{{\rho \atop \mu\nu}\right\} + u^{\rho}_{\mu\nu}.$$

The Christoffel symbols can be computed directly from $\gamma_{\mu\nu}$ as given by (5. 23). The non-vanishing ones are:

$$\left\{ {1 \atop 11} \right\} = \frac{\alpha'}{2\alpha}, \quad \left\{ {1 \atop 44} \right\} = \frac{\sigma'}{2\alpha'}, \quad \left\{ {1 \atop 33} \right\} = \left\{ {1 \atop 22} \right\} \sin^2\theta = \frac{-\beta'}{2\alpha} \sin^2\theta,$$

$$\left\{ {2 \atop 12} \right\} = \left\{ {3 \atop 13} \right\} = \frac{\beta'}{2\beta}, \quad \left\{ {2 \atop 33} \right\} = -\sin\theta\cos\theta, \quad \left\{ {3 \atop 23} \right\} = \frac{1}{\mathrm{tg}\theta},$$

$$\left\{ {4 \atop 14} \right\} = \frac{\sigma'}{2\sigma} \qquad \text{(IV. 19)}$$

$u^{\rho}_{\mu\nu}$ is computed from (3. 17)

$$u_{\mu\nu,\rho} = \underset{\mathrm{v}}{\Delta^{\sigma}_{\mu\rho}} \varphi_{\sigma\nu} + \underset{\mathrm{v}}{\Delta^{\sigma}_{\nu\rho}} \varphi_{\sigma\mu} \quad (u_{\mu\nu,\rho} = \gamma_{\rho\lambda} u^{\lambda}_{\mu\nu}).$$

From the values of $\varphi_{\mu\nu}$ in (5. 23), we see that only nine of the forty $u_{\mu\nu,\rho}$ do not vanish in the static case. These are:

$$u_{22,1} = 2\underset{\mathrm{v}}{\Delta^{3}_{12}} \varphi_{23}, \quad u_{33,1} = 2\underset{\mathrm{v}}{\Delta^{2}_{31}} \varphi_{23}, \quad u_{44,1} = -2\underset{\mathrm{v}}{\Delta^{1}_{14}} \varphi_{14},$$

$$u_{12,2} = -\underset{\mathrm{v}}{\Delta^{3}_{12}} \varphi_{23}, \quad u_{34,2} = -(\underset{\mathrm{v}}{\Delta^{1}_{23}} \varphi_{14} + \underset{\mathrm{v}}{\Delta^{2}_{24}} \varphi_{23}),$$

$$u_{13,3} = -\underset{\mathrm{v}}{\Delta^{2}_{31}} \varphi_{23}, \quad u_{24,3} = \underset{\mathrm{v}}{\Delta^{1}_{23}} \varphi_{14} + \underset{\mathrm{v}}{\Delta^{3}_{34}} \varphi_{23},$$

$$u_{23,4} = (\underset{\mathrm{v}}{\Delta^{2}_{24}} - \underset{\mathrm{v}}{\Delta^{3}_{34}}) \varphi_{23}, \quad u_{14,4} = \underset{\mathrm{v}}{\Delta^{1}_{14}} \varphi_{14}. \qquad \text{(IV. 20)}$$

Using IV. 18, we have

$$u^{1}_{33} = u^{1}_{22} \sin^2\theta = \frac{u}{\alpha} B_1 \sin\theta, \quad u^{1}_{44} = \frac{\sigma}{\alpha} \partial_1 \mathrm{Log}\left(1 - \frac{w^2}{\alpha\sigma}\right),$$

$$u^{2}_{12} = u^{3}_{13} = -\frac{u}{2\beta} B_1, \quad u^{2}_{34} = -u^{3}_{24} \sin^2\theta = \frac{w}{2\alpha} B_1 \sin\theta,$$

$$u^{4}_{14} = \frac{1}{2} \partial_1 \mathrm{Log}\left(1 - \frac{w^2}{\alpha\sigma}\right). \qquad \text{(IV. 21)}$$

Finally, the affine connection

$$\Delta^{\rho}_{\mu\nu} = \underset{\mathrm{v}}{\Delta^{\rho}_{\mu\nu}} + \left\{ {\rho \atop \mu\nu} \right\} + u^{\rho}_{\mu\nu}$$

is obtained by adding (IV 18), (IV. 19) and (IV. 21). We thus obtain the first column of table (5. 28). The non-static case leads to values of the affine connection given in the second column of (5. 28) in a completely similar way.

Appendix V

ISOTROPIC COORDINATE SYSTEM IN UNIFIED FIELD THEORY

The characteristic geodesics $f(x^{\mu}) = 0$ of the equations of the unified field theory are solutions of the partial differential equation ((10), p. 288).

$$\Delta_1 f \equiv g^{\mu\nu} \partial_{\mu} f \partial_{\nu} f \equiv h^{\mu\nu} \partial_{\mu} f \partial_{\nu} f = 0, \qquad (V.1)$$

where $g^{\mu\nu}$ has been split into symmetric and antisymmetric parts:

$$g^{\mu\nu} = h^{\mu\nu} + f^{\mu\nu},$$

$$g_{\mu\nu} = \gamma_{\mu\nu} + \varphi_{\mu\nu}.$$

The solutions of (V 1) are geodesics tangent to the cone $h_{\mu\nu} dx^{\mu} dx^{\nu} = 0$. They will also be wave fronts corresponding to $ds^2 = 0$ if the metric occurring in ds^2 coincides with $h^{\mu\nu}$.

Consider now the partial differential equations

$$g^{\mu\nu}(\underset{+}{\partial_{\mu}} f)_{;\nu} - M^{\rho} \partial_{\rho} f \equiv h^{\mu\nu}\partial_{\mu}\partial_{\nu} f - g^{\mu\nu}L^{\sigma}_{\mu\nu}\partial_{\sigma} f - M^{\rho}\partial_{\rho} f = 0, \qquad (V.2)$$

M^ρ being an arbitrary vector. (V. 2) generalizes the differential parameter of Beltrami

$$\Delta_2 f \equiv \left(g^{\underset{+-}{\mu\nu}} \partial_{\underset{+}{\mu}} f\right)_{;\nu} \equiv \left(g^{\underset{+-}{\mu\nu}} \partial_{\underset{-}{\nu}} f\right)_{;\mu} = M^\rho \partial_\rho f. \qquad (V. 3)$$

and admit $ds^2 = 0$ as the characteristic cones.

As in general Relativity, we can define geodesics

$$x^\rho = f(x'\mu) = \text{const.} \qquad (V. 4)$$

which are solutions of (V. 2). They are said to be isotropic. The choice of such geodesics can introduce great simplifications in the statements of the laws governing the fields.

From (V. 2), these geodesics must satisfy

$$g^{\mu\nu} L^\rho_{\mu\nu} = - M^\rho, \qquad (V. 5)$$

We define an isotropic system by the condition

$$g^{\mu\nu} L^\rho_{\mu\nu} = 0. \qquad (V. 6)$$

We use the equation (3.55)

$$L^\rho_{\mu\nu} = \left\{ {\rho \atop \mu\nu} \right\} + u^\rho_{\mu\nu} + \underset{V}{L^\rho_{\mu\nu}}, \quad (L_\rho = 0)$$

with

$$u^\rho_{\mu\nu} = - \gamma^{\rho\sigma} (\underset{V}{L_{\mu\sigma}}, _{\bar\nu} + \underset{V}{L_{\nu\sigma}}, _{\bar\mu}) = -\gamma^{\rho\sigma} (\varphi_{\nu\lambda} \underset{V}{L^\lambda_{\mu\sigma}} + \varphi_{\mu\lambda} \underset{V}{L^\lambda_{\nu\sigma}}). \qquad (3. 17)$$

Use of the above with (1. 18) yields

$$g^{\mu\nu} L^\rho_{\mu\nu} = h^{\mu\nu} \left\{ {\rho \atop \mu\nu} \right\} - \gamma^{\rho\sigma} \left(\frac{\gamma}{g} \gamma^{\mu\nu} + \frac{\varphi}{g} \varphi^{\mu\lambda} \varphi^{\nu\tau} \gamma_{\lambda\tau} \right)$$

$$\times (\varphi_{\nu\epsilon} \underset{V}{L^\epsilon_{\mu\sigma}} + \varphi_{\mu\epsilon} \underset{V}{L^\epsilon_{\nu\sigma}}) \qquad (V. 7)$$

$$+ \left(\frac{\varphi}{g} \varphi^{\mu\nu} + \frac{\gamma}{g} \gamma^{\mu\lambda} \gamma^{\nu\sigma} \varphi_{\lambda\sigma} \right) \underset{V}{L^\rho_{\mu\nu}} = 0.$$

The computation of these expressions does not require a complete knowledge of $L^{\rho}_{\mu\nu}$ but simply knowledge of the quantities

$$B_\rho = \frac{1}{2}\varphi^{\mu\nu}\underset{V}{L_{\mu\nu,\rho}}, \qquad A_\rho = \frac{1}{2}\gamma^{\mu\lambda}\gamma^{\nu\tau}\varphi_{\lambda\tau}\underset{V}{L_{\mu\nu,\rho}}$$

which we introduced in the determination of the general solution of $g_{\underset{+}{\mu}\underset{-}{\nu};\rho} = 0$ (cf. (3. 30) and (3. 31). We then obtain from (V. 7)

$$h_{\mu\nu}\left\{{\rho \atop \mu\nu}\right\} + \frac{1}{2}\gamma^{\rho\sigma} f^{\mu\nu}\varphi_{\mu\sigma\nu} = 0. \qquad (V. 8)$$

Making the approximations

$$\gamma_{\mu\nu} = \eta_{\mu\nu} + \epsilon\underset{1}{\gamma}_{\mu\nu} + \epsilon^2\underset{2}{\gamma}_{\mu\nu} + \dots,$$

$$\varphi_{\mu\nu} = \underset{1}{\varphi}_{\mu\nu} + \underset{2}{\varphi}_{\mu\nu} + \dots,$$

(V. 8) reduces, to first order:

$$\frac{1}{2}\gamma\gamma^{\mu\nu}\left\{{\rho \atop \mu\nu}\right\} = \eta^{\mu\nu}\eta^{\rho\sigma}\left(\partial_\mu\underset{1}{\gamma}_{\nu\sigma} - \frac{1}{2}\partial_\sigma\underset{1}{\gamma}_{\mu\nu}\right) = 0 \qquad (V. 9)$$

or

$$\epsilon_\mu\left(\partial_\mu\underset{1}{\gamma}_{\mu\rho} - \frac{1}{2}\partial_\rho\underset{1}{\gamma}_{\mu\mu}\right) = 0.$$

This is the classical expression resulting from the choice of an isotropic system in general Relativity. This is the choice adopted by Schrödinger in defining transformations of an appropriate system of reference (cf. (4. 56) p. 60). To second order, we have

$$\eta^{\mu\nu}\left(\partial_\mu\underset{2}{\gamma}_{\nu\rho} - \frac{1}{2}\partial_\rho\underset{2}{\gamma}_{\mu\nu}\right) + \underset{1}{\gamma}^{\mu\nu}\left(\partial_\mu\underset{1}{\gamma}_{\nu\rho} - \frac{1}{2}\partial_\rho\underset{1}{\gamma}_{\mu\nu}\right)$$

$$+ \frac{1}{2}\eta^{\mu\lambda}\eta^{\nu\tau}\underset{1}{\varphi}_{\lambda\tau}\underset{1}{\varphi}_{\mu\rho\nu} = 0, \qquad (V. 10)$$

that is, if we assume $\underset{1}{\gamma}_{\mu\nu} = 0$ (cf. Chapter 4, § 5)

$$\epsilon_\mu\left(\partial_\mu\underset{2}{\gamma}_{\mu\rho} - \frac{1}{2}\partial_\rho\underset{2}{\gamma}_{\mu\mu}\right) + \frac{1}{2}\epsilon_\mu\epsilon_\nu\underset{1}{\varphi}_{\mu\nu}\underset{1}{\varphi}_{\mu\rho\nu} = 0.$$

References

1. Bergmann: An Introduction to the Theory of Relativity Prentice-Hall, Inc., New York, 1942.
2. Cartan: Soc. Math., **1**, 141, 1922.
3. Cartan: Ann. Sc. Ec. Norm. Sup., vol. **40**, 1923.
4. Chazy: La théorie de la Relativité et la Mécanique Céleste. Gauthier-Villars, Paris, 1928.
5. Eddington: The Mathematical Theory of Relativity. Cambridge University Press, London, 1954.
6. Einstein: Théorie de la gravitation généralisée (traduction Solovine, Gauthier-Villars, 1951).
7. Eisenhart: Non Riemannian Geometry, American Mathematical Society, New York 1927.
8. Eisenhart: Proc. Nat. Acad. Sc. U. S. A. **37**, -311, 1951; **38**, 505, 1952; **39**, 546, 1953.
9. Laue: Theory of Relativity. Gauthier-Villars, 1924-1926.
10. Lichnerowicz: Les théories relativistes de la gravitation et de l'électromagnétisme (Cours au Collège de France, Masson 1955).
11. Möller: The Theory of Relativity, Oxford University Press 1952.
12. Tonnelat: Les Theories Unitaires de L'electromagnetisme et de la Gravitation (Gauthier-Villars).
13. Weyl: Time, Space and Gravitation
14. Einstein: The Meaning of Relativity, (Appendix II) Princeton University Press, 3rd edition 1950.

15. Einstein: The Meaning of Relativity (Appendix) 4th edition, Princeton University Press 1953.
16. Einstein and Strauss: Journ. Math. **47**, 731, 1946.
17. Einstein: Ann. Math. (Princeton) **46**, 578, 1945; **47**, 731, 1946. Rev. Mod. Phys., **20**, 35, 1948; **21**, 343, 1949.
18. Einstein and Kaufman: Louis de Broglie, Physicien et Penseur, Albin-Michel, Paris 1952.
19. Einstein: Ibid.
20. Einstein: Phys. Rev. **89**, 321, 1953.
21. Einstein and Kaufman: Ann. Math., **59**, 230, 1954.
22. Schrödinger: Proc. Roy. Ir. Acad., **49A**, 43, 1943; **49A**, 135, 1943; **49A**, 225, 237, 275, 1944.
23. Schrödinger: Ibid, **50A**, 143, 1945; **50A**, 223, 1945.
24. Schrödinger: Ibid, **51**, 41, 1945.
25. Schrödinger: Ibid, **51A**, 163, 1947; **51**, 205, 1948; **52**, 1, 1948.
26. Schrödinger: Ibid, **56A**, 13, 1954.
27. Bandyopadhyay: Phys. Rev. **89**, 1161, 1953.
28. Bose: C. R. Acad. Sc., **236**, 1333, 1953.
29. Bose: J. Phys. Rad., **14**, 641, 1953.
30. Bose: Ann. Math., **59**, 171, 1952.
31. Cap. Acta Phys. Austr., **6**, 135, 1952.
32. Finzi: Atti Accad. Nazion, Lincei, **14**, 591, 1953; Ric. Sc. ital., **20**, 1901, 1950; Univ. Roma, Rendic, Mat., **11**, 75, 1952.
33. Hely: C. R. Acad. Sc., **239**, 385, 1954.
34. Hlavaty: Proc. Nat. Acad. Sc. (U. S. A.) **39**, 243, 1952; J. Rational Mechanics and Analysis, **2**, 1, 1953; **3**, 103, 1954.
35. Hlavaty: Proc. Nat. Acad. Sc. (U. S. A.), **39**, 507, 1953. Geometry of Einstein's Unified Field Theory, Noordhoff Ltd., Groninger, Holland 1957.
36. Hoffmann: Bull. Amer. Phys. Soc. **23**, 5 and 44, 1948; Phys. Rev., 73, 30, 1948; **73**, 531, 1948; **73**, 1042, 1948.
37. Johnson: Phys. Rev. **89**, 320, 1953.
38. Jordan: Z. Phys., **124**, 602, 1948.
39. Kursunoglu: Phys. Rev., **88**, 1369, 1952.

40. Lenoir: C. R. Acad. Sc., **237**, 424, 1953.
41. Lichnerowicz: C. R. Acad. Sc., **237**, 1383, 1953.
42. Pastori: 1st. Lombardo Sc. Lettere, Rendic., Cl. Sc. mat. nat., **84**, 509, 1951.
43. Winogradzki: C. R. Acad. Sc. **239**, 1359, 1954; **240**, 945, 1955; J. Phys. Rad., **16**, 1955.
44. Clauser: Atti. Accad. Nazion, Lincei, **15**, 171, 1953.
45. Freistadt: Bull. Amer. Phys. Soc., **25**, 28, 1950.
46. Hlavaty: Proc. Nat. Acad. Sc. (U. S A.), **38**, 415, 1952.
47. Hlavaty: Ibid, **38**, 1052, 1952.
48. Ikeda: Progr. Theor. Phys., Japan, **7**, 127, 1952.
49. Ingraham: Ann. Math., **52**, 743, 1950.
50. Kursunoglu: Phys. Rev. **82**, 289, 1951.
51. Pastori, Atti. Accad. Nazion, Lincei, **12**, 302, 1952.
52. Strauss: Rev. Mod. Phys., **21**, 414, 1949.
53. Tonnelat: C R. Acad. Sc., **230**, 182, 1950; **231**, 470, 1950; **231**, 487, 1950; **231**, 512, 1950; **232**, 2407, 1951.
54. Tonnelat: J. Phys. Rad. **12**, 81, 1951.
55. Tonnelat: J. Phys. Rad., **13**, 177, 1952.
56. Tonnelat: C. R. Accad. Sc., **239**, 1468, 1954.
57. Tonnelat: J. Phys. Rad., **16**, 21, 1955.
58. Udeschini: Atti Accad. Lincei, Rendic., **10**, 390, 1951.
59. Udeschini: Ibid, **9**, 256, 1950; **10**, 21, 1951.
60. Udeschini: Ibid, **15**, 165, 1953.
61. Bandyopadhyay: Nature, **167**, 648, 1951.
62. Bandyopadhyay: Indian J. Phys., **25**, 257, 1951.
63. Bonnor: Proc. Roy. Soc. **209**, 353, 1951; **210**, 427, 1952.
64. Clark: Phil. Mag. **39**, 747, 1948.
65. Mavrides: C. R Acad. Sc., **238**, 1566, 1954; **239**, 637, 1954; **238**, 1643, 1954. J. Phys. Rad., **16**, 1955.
66. Mavridès: C. R. Acad. Sc., **239**, 1597, 1954.
67. Mineur: Bull. Soc. Math. Fr., **56**, 50, 1928.
68. Narlikar and Karmakar: Nature, **157**, 515, 1946.
69. Narlikar and Vaidya: Proc. Nat. Inst. Sc. India, **14**, 53, 1948.
70. Papapetrou: Proc. Roy. Ir. Acad., **51A**, 191, 1947.
71. Papapetrou: Ibid, **52A**, 69, 1948.

72. Papapetrou: Nature, **168**, 40, 1951.
73. Rickaysen and Kursunoglu: Phys. Rev., **89**, 522, 1953.
74. Rosen: Phys. Rev. **70**, 565, 1946.
75. Schrödinger: Proc. Roy. Ir. Acad., **49A**, 225, 1944; Nature, **153**, 572, 1944.
76. Schrödinger and Papapetrou: Nature, **168**, 40, 1951.
77. Takeno, Ikeda and Abe: Progr. Theor. Phys., Japan, **6**, 837, 1951.
78. Tonnelat: C. R. Acad. Sc. **239**, 231, 1954.
79. Vaidya: Phys. Rev., **83**, 10, 1951.
80. Vaidya: Curr. Sc. India, **21**, 96, 1952.
81. Vaidya: J. Univ. Bombay, **21A**, 1, 1952.
82. Vaidya: Indian. Sc. Congress Asso., Calcutta 1953; Phys., Rev., **90**, 695, 1953.
83. Vaidya: Phys. Rev. **96**, 5, 1954.
84. Wyman: Phys. Rev. **70**, 74, 1946.
85. Wyman: Canad. J. Math., **2**, 427, 1950.
86. Bergmann: Phys. Rev., **90**, 315, 1953.
87. Callaway: Phys. Rev. **92**, 1567, 1953.
88. Darmois: Mem. S. Math., vol. **25**, 1927.
89. Egorov: Dokl. Akad. Nauk, U. S. S. R., **87**, 693, 1952.
90. Einstein and Grommer: Sitzb. Berl. Akad., vol **2**, 1927.
91. Einstein: Canad. J. Math., **2**, 120, 1950.
92. Einstein, Infeld and Hoffmann: Ann. Math., **39**, 65, 1938.
93. Einstein and Infeld: Ann. Math., **41**, 455, 1940.
94. Einstein and Infeld: Canad. J Math., **1**, 209, 1949.
95. Einstein: Revista-Univ. Nat. de Tucuman, A_2, vol **11**, 1941.
96. Einstein and Pauli: Ann. Math., **44**, 131, 1945.
97. Fock: Journ. Phys. U. S. S. R., **1**, 81, 1939.
98. Goldberg: Phys. Rev., **89**, 263, 1953.
99. Goldberg and Schiller: Bull. Amer. Phys. Soc., **27**, 45, 1952.
100. Haywood: Proc. Phys. Soc., **65A**, 170, 1952.
101. Infeld: Nature, **166**, 1075, 1950.
102. Infeld: Acta Physica Polonica, **10**, 284, 1950.
103. Infeld: Canad. J. Math., **5**, 17, 1953.

104. Infeld and Wallace: Phys. Rev., **57**, 797, 1940.
105. Infeld and Schild: Rev. Mod. Phys., **21**, 408, 1949.
106. Infeld and Scheidegger: Canad. J. Math., **3**, 195, 1951.
107. Kursunoglu: Proc. Phys. Soc. London, **65**, 81, 1952.
108. Lanczos: Proc. Ind. Acad. Sc., **50**, 170, 1941.
109. Laurikainen: Ann. Acad. Sc. feunicae, **1A**, 45, 1950.
110. Melvin: Bull. Amer. Phys. Soc. **28**, 43, 1953.
111. Mira Fernandes: Univ. Lisboa, Rev. Fac. Ci., 2nd series **1A**, 173, 1950.
112. Papapetrou: Phys. Rev., **73**, 1105, 1948.
113. Papapetrou: Proc. Roy. Soc., London **64**, 57, 1951; **64**, 302, 1951.
114. Schrödinger: Proc. Roy. Ir. Acad., **51A**, 147, 1947.
115. Schrödinger: Proc. Roy. Ir. Acad., **52A**, 1, 1948.
116. Schrödinger: Proc. Roy. Ir. Acad., **54A**, 79, 1951.
117. Schrödinger: Comm. Dublin Inst., A No. 6, 1951.
118. Winogradzki: C. R. Acad. Sc., **238**, 996, 1954.
119. Arley and Fuchs: Nature, **161**, 598, 1948.
120. Blackett: Magnetic Field of Rotating Massive Bodies, (VIII Conseil de Physique Solvay, Sept. 27-Oct. 2, 1948).
121. Born: Proc. Roy. Soc., London, **143A**, 410, 1934.
122. Born and Infeld: Proc. Roy. Soc., London, **144A**, 425, 1934.
123. Callaway: Phys. Rev., **96**, 778, 1954.
124. Castoldi: Atti. Accad. ligure Sc. Lettere, 7, 307, 1951.
125. Einstein: Berlin Sitzungsberichte, pp. 32-76, 1923.
126. Fourès-Bruhat: Acta. Math., **88**, pp. 141-225, 1952.
127. Hittmair and Schrödinger: Comm. Dublin Inst., A, No. 8, 1951.
128. Luchak: Canad. J. Phys., **29**, 470, 1951.
129. Narlikar: Phys. Rev., **76**, 868, 1949; Rev. Mod. Phys., **20**, 35, 1948.
130. Santalo: Revista Mat. y Fis. Teor. Tucamun, p. 19, 1954.

Index